PURE MATHEMATICS

5. CARTESIAN AND POLAR CURVE SKETCHING

Third Edition
By
Anthony Nicolaides

P.A.S.S. PUBLICATIONS

Private Academic & Scientific Studies Limited

© A. NICOLAIDES 1991, 1995, 2007

First Published in Great Britain 1991 by
Private Academic & Scientific Studies Limited

ISBN–13   978–1–872684–63–5

THIRD EDITION 2008

This book is copyright under the Berne Convention.
All rights are reserved. Apart as permitted under the Copyright Act, 1956, no part of this publication may be reproduced, stored in a retrieval system, or transmitted in any form of by any means, electronic, electrical, mechanical, optical, photocopying, recording or otherwise, without the prior permission of the publisher.

Titles by the same author.

Revised and Enhanced

1. Algebra.   GCE A Level   ISBN–13   978–1–872684–82–6   £11–95

2. Trigonometry.   GCE A Level   ISBN–13   978–1–872684–87–1   £11–95

3. Complex Numbers.   GCE A Level   ISBN–13   978–1–872684–92–5   £9–95

4. Differential Calculus and Applications.   GCE A Level   ISBN–13   978–1–872684–97–0   £9–95

5. Cartesian and Polar Curve Sketching.   GCE A Level   ISBN–13   978–1–872684–63–5   £9–95

6. Coordinate Geometry in two Dimensions.   GCE A Level   ISBN–13   978–1–872684–68–0   £9–95

7. Integral Calculus and Applications.   GCE A Level   ISBN–13   978–1–872684–73–4   £14–95

8. Vectors in two and three dimensions.   GCE A Level   ISBN–13   978–1–872684–15–4   £9–95

9. Determinants and Matrices.   GCE A Level   ISBN–13   978–1–872684–16–1   £9–95

10. Probabilities.   GCE A Level   ISBN–13   978–1–872684–17–8   £8–95
    This book includes the full solutions

11. Success in Pure Mathematics: The complete works of GCE A Level. (1–9 above inclusive)   ISBN   978–1–872684–93–2   £39–95

12. Electrical & Electronic Principles. First year Degree Level   ISBN–13 978–1–872684–98–7   £16–95

13. GCSE Mathematics Higher Tier Third Edition.   ISBN–13   978–1–872684–69–7   £19–95

All the books have answers and a CD is attached with FULL SOLUTIONS of all the exercises set at the end of the book.

Preface

This book, which is part of the GCE A level series in Pure Mathematics covers the specialized topic of Cartesian and Polar Curve Sketching.

The GCE A level series success in Pure Mathematics is comprised of nine books, covering the syllabuses of most examining boards. The books are designed to assist the student wishing to master the subject of Pure Mathematics. The series is easy to follow with minimum help. It can be easily adopted by a student who wishes to study it in the comforts of his home at his pace without having to attend classes formally; it is ideal for the working person who wishes to enhance his knowledge and qualification. Cartesian and Polar Curve Sketching book, like all the books in the series, the theory is comprehensively dealt with, together with many worked examples and exercises. A step by step approach is adopted in all the worked examples. A CD is attached to the book with FULL SOLUTIONS of all the exercises set at the end of each chapter.

This book develops the basic concepts and skills that are essential for the GCE A level in Pure Mathematics.

The Cartesian and Polar Curve Sketching book covers thoroughly the C1 and FP1 topics in the recent syllabus.

A. Nicolaides

SUCCESS IN PURE MATHEMATICS
5. CARTESIAN AND POLAR CURVE SKETCHING

CARTESIAN CURVE SKETCHING AND APPLICATIONS

5. CARTESIAN AND POLAR CURVE SKETCHING

CONTENTS

CARTESIAN CURVE SKETCHING AND APPLICATIONS

1.	**QUADRATIC AND CUBIC FUNCTIONS**	3
	Cubic functions	4
	To prove the maximum power transfer theorem	6
	Exercises 1	9
2.	**SKETCH THE ALGEBRAIC–EXPONENTIAL COMPOSITE FUNCTIONS**	11
	Exercises 2	13
3.	**ALGEBRAIC FUNCTIONS WITH ASYMPTOTES**	14
	Exercises 3	17
4.	**SIMPLE TRANSFORMATIONS**	18
	Exercises 4	25

POLAR CURVE SKETCHING AND APPLICATIONS

5.	**POLAR COORDINATES SYSTEM**	27
	Exercises 5	28
6.	**RELATIONSHIP BETWEEN POLAR AND CARTESIAN COORDINATES**	29
	Exercises 6	29
7.	**HALF-LINES OR PART-LINES**	31
	Exercises 7	31
8.	**THE POLAR EQUATIONS OF STRAIGHT LINES**	32
	Exercises 8	33
9.	**POLAR CURVE SKETCHING**	34
	Exercises 9	36
10.	**AREA OF THE TRIANGLE** OPQ **WHERE** $P(r_1, \Theta_1)$ **AND** $Q(r_2, \Theta_2)$	37
	Exercises 10	38
11.	**POLAR TO CARTESIAN AND VICE-VERSA**	39
	To convert cartesian to polar	39

12.	**THE STATIONARY VALUES OF** $r \sin \Theta$	42
	Exercises 12	43
13.	**THE STATIONARY VALUES OF** $r \cos \Theta$	44
	Exercises 13	44
14.	**AREAS OF POLAR CURVES**	45
	Exercises 14	46
	MISCELLANEOUS	47
	ANSWERS	53
	INDEX	57

CARTESIAN CURVE SKETCHING AND APPLICATIONS

1

Quadratic and Cubic Functions

Turning points or stationary points or maximum and minimum points, Points of inflection

$y = f(x) = ax^2 + bx + c$.

A quadratic function is a parabola which either has a maximum or a minimum. The maximum or minimum can be easily determined by examining the second derivative, if $\frac{d^2y}{dx^2} > 0$ there is a minimum and if $\frac{d^2y}{dx^2} < 0$ there is a maximum.

The first derivative, however, determines at which point does the maximum or minimum occurs. $\frac{dy}{dx} = f'(x) = 2ax + b$ is the gradient at any point of the function, when $\frac{dy}{dx} = 0$, the turning or stationary point is determined

$2ax + b = 0$

$$\boxed{x = -\frac{b}{2a}}.$$

To sketch the graph $y = ax^2 + bx + c$.

(a) If $x = 0$, $y = c$

(b) At $x = -\frac{b}{2a}$ there is a maximum or minimum, or a turning point or a stationary point.

(c) $\frac{d^2y}{dx^2} > 0$ gives a minimum, $\frac{d^2y}{dx^2} < 0$ gives a maximum.

The graph is then easily sketched.

WORKED EXAMPLE 1

Sketch the parabola $y = f(x) = x^2 - 2x - 8$, indicating the coordinates where the curve intersects x and y axes.

Solution 1

(a) If $x = 0$, $y = -8$, $f(0) = -8$

(b) when $x = -\frac{b}{2a} = \frac{-(-2)}{2(1)} = 1$

(c) $\frac{dy}{dx} = 2x - 2$, $\frac{d^2y}{dx^2} = 2 > 0$ a minimum

$f(1) = y_{min} = 1^2 - 2 - 8 = -9$

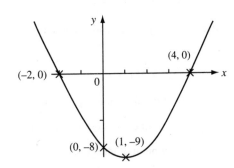

Fig. 5-I/1

when $y = 0$, $x = \frac{2 \pm \sqrt{4 + 32}}{2} = \frac{2 \pm 6}{2}$, $x = 4$, $x = -2$.

WORKED EXAMPLE 2

Sketch the quadratic function $y = f(x) = -x^2 - 2x + 3$, indicating the coordinates where the curve intersects the x and y axes.

Solution 2

(a) $f(0) = 3$

(b) $f'(x) = -2x - 2 = 0$, $x = -1$

3

(c) $f''(x) = -2 < 0$, a maximum

$f(-1) = -(-1)^2 - 2(-1) + 3 = -1 + 2 + 3 = 4$

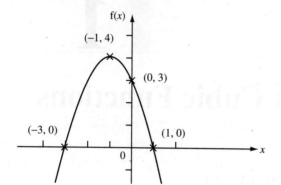

Fig. 5-I/2

$f(x) = 0, x = \dfrac{2 \pm \sqrt{4+12}}{-2} = \dfrac{2 \pm 4}{-2}, x = -3, x = 1.$

Cubic Functions

$y = f(x) = ax^3 + bx^2 + cx + d.$

Procedure for sketching the cubic function.

(a) $f(0) = d$

(b) $f'(x) = 3ax^2 + 2bx + c$

$f'(x) = 0$ for turning points x_1, x_2 if they exist

(c) $f''(x_1) > 0$ minimum $f''(x_2) < 0$ maximum

(d) $f''(x) = 0$ for point of inflection.

WORKED EXAMPLE 3

Sketch the graph $y = f(x) = x^3 - 2x^2 - 5x + 6.$

Solution 3

(a) $f(0) = 6$

(b) $f'(x) = 3x^2 - 4x - 5$

$f'(x) = 0$ for turning points

$3x^2 - 4x - 5 = 0$

$x = \dfrac{4 \pm \sqrt{16+60}}{6} = \dfrac{4 \pm \sqrt{76}}{6}$

$x_1 = 2.120, x_2 = -0.786$

(c) $f''(x) = 6x - 4$

$f''(2.120) = 6(2.120) - 4 = 8.72 > 0$ min.

$f''(-0.786) = 6(-0.786) - 4 = -8.72 < 0$ max.

(d) $f''(x) = 6x - 4 = 0, x = \dfrac{2}{3}$ point of inflection.

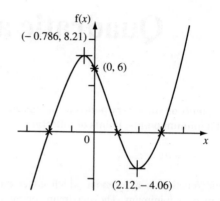

Fig. 5-I/3

$f(2.120) = (2.120)^3 - 2(2.120)^2 - 5(2.120) + 6$
$= -4.06$

$f(-0.786) = (-0.786)^3 - 2(-0.786)^2$
$-5(-0.786) + 6 = +8.21.$

Point of Inflection

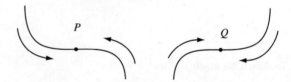

Fig. 5-I/4

At the points P and Q, the rate of change of the gradient $= \dfrac{d}{dx}\left(\dfrac{dy}{dx}\right) = 0$, there is neither a maximum nor a minimum.

WORKED EXAMPLE 4

Determine the nature of the turning points of the function and investigate the point of inflection $y = f(x) = x^3 + x^2 + 3x + 3.$

Sketch the graph.

Solution 4

(a) $f(0) = 3$

(b) $f'(x) = 3x^2 + 2x + 3$,

$f'(x) \neq 0$ since x has complex roots, there are no turning points.

(c) $f''(x) = 6x + 2$

$f''(x) = 0$ when $x = -\frac{1}{3}$ point of inflection.

(d) $f(-1) = (-1)^3 + (-1)^2 + 3(-1) + 3$

$\quad = -1 + 1 - 3 + 3 = 0$

$x + 1$ is a factor of $x^3 + x^2 + 3x + 3$

$$\begin{array}{r} x^2 + 3 \\ x+1\overline{\smash{\big)}\,x^3 + x^2 + 3x + 3} \\ \underline{x^3 + x^2} \\ 3x + 3 \\ \underline{3x + 3} \\ 0 \end{array}$$

$f(x) = (x+1)(x^2 + 3)$

$f(-1) = 0$, $x = -1$ intersect the x-axis at $x = -1$ and the y-axis at $y = 3$.

when $x = -\frac{1}{3}$,

$f\left(-\frac{1}{3}\right) = \left(-\frac{1}{3}\right)^3 + \left(-\frac{1}{3}\right)^2 + 3\left(-\frac{1}{3}\right) + 3 = 2.$

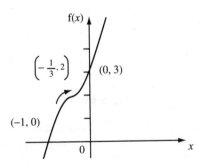

Fig. 5-I/5

WORKED EXAMPLE 5

A cardboard of 1 square metre is given and it is required to construct an open box of maximum volume by cutting from each corner equal squares, and the ends are turned up as shown. Determine the height of the box and the maximum volume in metres. Fig. 5-I/6 and Fig. 5-I/7 show the cardboard and open box respectively.

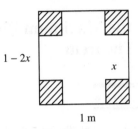

Fig. 5-I/6

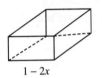

Fig. 5-I/7

Solution 5

$V = (1 - 2x)^2 x \qquad V = (1 - 4x + 4x^2)x$

$\qquad\qquad\qquad\qquad\quad = x - 4x^2 + 4x^3$

$V = 4x^3 - 4x^2 + x \qquad \dfrac{dV}{dx} = 12x^2 - 8x + 1.$

For maximum volume, $\dfrac{dV}{dx} = 0$

$12x^2 - 8x + 1 = 0$

$x = \dfrac{8 \pm \sqrt{64 - 48}}{24} = \dfrac{8 \pm 4}{24}$

$x = \dfrac{1}{2}$ m or $x = \dfrac{1}{6}$ m.

Obviously, if we take $x = \frac{1}{2}$, there will be nothing left, $x = \frac{1}{6}$ m is the answer for maximum volume. This can be checked by obtaining the second derivative

$\dfrac{d^2V}{dx^2} = 24x - 8.$

If $x = \dfrac{1}{6}$, $\dfrac{d^2V}{dx^2} = 24\left(\dfrac{1}{6}\right) - 8 = -4$ it is maximum

$V_{max} = 4\left(\dfrac{1}{6}\right)^3 - 4\left(\dfrac{1}{6}\right)^2 + \dfrac{1}{6} = \dfrac{4}{216} - \dfrac{4}{36} + \dfrac{1}{6}$

$\qquad = \dfrac{4 - 24 + 36}{216}$

$V_{\max} = \dfrac{16}{216} = \dfrac{2}{27}$

$x = \dfrac{1}{6}$ m and $V_{\max} = \dfrac{2}{27}$ m^3.

To Prove the Maximum Power Transfer Theorem

WORKED EXAMPLE 6

A source of e.m.f., E, and internal resistance, r, is connected across a variable load, R. Determine the condition between the internal resistance, r, and the external load, R, so that maximum power is dissipated in R.

Solution 6

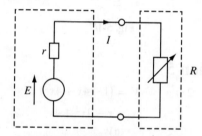

Fig. 5-I/8

The current through R is given $I = \dfrac{E}{r+R}$ the power dissipated in R is given $P = I^2 R$

$P = \left(\dfrac{E}{r+R}\right)^2 R = \dfrac{E^2 R}{(r+R)^2}$ in this expression P and R are the two variables, E and r are constants.

It is required to find the rate of change of the power, P, with respect to the resistance, R.

Using the quotient rule

$\dfrac{dP}{dR} = \dfrac{E^2(r+R)^2 - 2(r+R)E^2 R}{(r+R)^4}$ dividing numerator and denominator by $r+R$ we have

$\dfrac{dP}{dR} = \dfrac{E^2(r+R) - 2E^2 R}{(r+R)^3}$ for maximum power

$\dfrac{dP}{dR} = 0 \quad \dfrac{E^2(r+R) - 2E^2 R}{(r+R)^3} = 0,$

$E^2(r+R) - 2E^2 R = 0 \quad (r+R) - 2R = 0,$

$r + R = 2R \qquad R = r$

$P_{\max} = \dfrac{E^2 r}{(r+r)^2} = \dfrac{E^2 r}{4r^2} = \dfrac{E^2}{4r} \qquad P_{\max} = \dfrac{E^2}{4r}.$

The second derivative must be negative

$\dfrac{d^2 P}{dR^2} = \dfrac{(E^2 - 2E^2)(r+R)^3 - [(r+R)E^2 - 2E^2 R] \cdot 3(r+R)^2}{(r+R)^6}$

simplifying

$\dfrac{d^2 P}{dR^2} = \dfrac{-E^2(r+R) - 3E^2(r+R) + 6E^2 R}{(r+R)^4}$

$= \dfrac{-4E^2(r+R) + 6E^2 R}{(r+R)^4} = \dfrac{-4E^2 r + 2E^2 R}{(r+R)^4}$

$= -\dfrac{2E^2 r}{(r+R)^4} = -\dfrac{2E^2 r}{2^4 r^4} = -\dfrac{E^2}{8r^3} < 0$ if $R = r$.

WORKED EXAMPLE 7

A cylindrical container, closed at both ends, contains a volume 24π m^3.

Given that the total external surface area is a minimum. Calculate:

(i) the base radius
(ii) the height
(iii) the minimum surface area.

Solution 7

The surface area is given $S = 2\pi r^2 + 2\pi r h$ where r is the base radius and h is its height.

Fig. 5-I/9

The volume of the cylindrical container $V = \pi r^2 h$, $\pi r^2 h = 24\pi$, $h = \dfrac{24}{r^2}$ the volume was given so that we know the relationship between its height, h, and base radius, r.

$S = 2\pi r^2 + 2\pi r h$ substituting $h = \dfrac{24}{r^2}$ in this equation

$S = 2\pi r^2 + 2\pi r \cdot \dfrac{24}{r^2}$, $S = 2\pi r^2 + \dfrac{48\pi}{r} = 2\pi r^2 + 48\pi r^{-1}$, $S = 2\pi r^2 + 48\pi r^{-1}$ differentiating S with respect to r

$\dfrac{dS}{dr} = 4\pi r + (-1)48\pi r^{-2}$

$\dfrac{dS}{dr} = 4\pi r - \dfrac{48\pi}{r^2} = 4\pi r - 48\pi r^{-2}$.

For maximum or minimum surface area $\dfrac{dS}{dr} = 0$

$4\pi r - \dfrac{48\pi}{r^2} = 0 \qquad r^3 = 12$

$r = \sqrt[3]{12} = 12^{\frac{1}{3}} = 2.29$ m.

The second derivative $\dfrac{d^2S}{dr^2} = 4\pi - 48(-2)\pi r^{-3} = 4\pi + \dfrac{96\pi}{r^3}$ this is certainly positive, therefore the surface area is minimum for $r = \sqrt[3]{12}$ and

$h = \dfrac{24}{r^2} = \dfrac{24}{\left(12^{\frac{1}{3}}\right)^2} = \dfrac{24}{12^{\frac{2}{3}}} \qquad h = 4.58$ m

$S_{\min} = 2\pi r^2 + 2\pi r h$

$= 2\pi \left(12^{\frac{1}{3}}\right)^2 + 2\pi \left(12^{\frac{1}{3}}\right)(4.58)$

$= 32.93 + 65.9 = 98.85 = 98.8$ m².

WORKED EXAMPLE 8

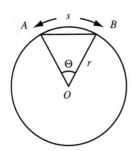

Fig. 5-I/10

The length of the arc s is given by $s = r\Theta$.

The area of the triangle ABO $\Delta_1 = \dfrac{1}{2}r^2 \sin \Theta$.

The area of the sector ABO is $\Delta_2 = \dfrac{1}{2}r^2 \Theta$.

Express the area of the triangle and sector in terms of s.

Find the maximum area of the triangle Δ_1 assuming r is constant and s is variable. Verify that the area is maximum.

What is the area of the sector, Δ_2, for this condition.

Quadratic and Cubic Functions — 7

Solution 8

$\Delta_1 = \dfrac{1}{2}r^2 \sin \Theta = \dfrac{1}{2}r^2 \sin \left(\dfrac{s}{r}\right)$ for maximum area

$\dfrac{d\Delta_1}{ds} = 0$

$\dfrac{d\Delta_1}{ds} = \dfrac{1}{2}r^2 \left(\dfrac{1}{r}\right) \cos \left(\dfrac{s}{r}\right) = 0 \qquad \dfrac{1}{2}r \cos \dfrac{s}{r} = 0$

or $\cos \dfrac{s}{r} = 0$, but $\cos \dfrac{\pi}{2} = 0 \qquad \cos \dfrac{s}{r} = \cos \dfrac{\pi}{2}$

therefore $s = r\dfrac{\pi}{2}$ the second derivative

$\dfrac{d^2\Delta_1}{ds^2} = -\dfrac{1}{2}r^2 \left(\dfrac{1}{r}\right)\left(\dfrac{1}{r}\right) \sin \dfrac{s}{r}$

or $\dfrac{d^2\Delta_1}{ds^2} = -\dfrac{1}{2} \sin \dfrac{s}{r}$

$= -\dfrac{1}{2} \sin r \dfrac{\pi}{2} \dfrac{1}{r}$

$= -\dfrac{1}{2} \sin \dfrac{\pi}{2} = -\dfrac{1}{2}$

which is, negative and indicates that Δ_1 is a maximum

$(\Delta_1)_{\max} = \dfrac{1}{2}r^2 \sin \left(\dfrac{s}{r}\right) = \dfrac{1}{2}r^2 \sin \left(r\dfrac{\frac{\pi}{2}}{r}\right) = \dfrac{1}{2}r^2$.

$\Delta_2 = \dfrac{1}{2}r^2 \Theta = \dfrac{1}{2}r^2 \dfrac{s}{r} = \dfrac{1}{2}rs = \dfrac{1}{2}rr\dfrac{\pi}{2} = \dfrac{r^2 \pi}{4}$.

WORKED EXAMPLE 9

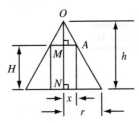

Fig. 5-I/11

A right circular cylinder is to be cut from right circular solid cone as shown. If x is the radius of the cylinder, find an expression for the volume, V, of the cylinder in terms of x, r and h, hence find the maximum volume of the cylinder as x varies, if $r = 3$ m and $h = \dfrac{3}{4\pi}$ m.

Solution 9

The volume of the cylinder is given $V = \pi x^2 \times$ height. The height of the cylinder is found from the similar right angled triangles OMA and ONB from which we have
$\frac{x}{r} = \frac{OM}{ON} = \frac{h - \text{height of cylinder}}{h}$

$\frac{x}{r} = \frac{h}{h} - \frac{\text{height of the cylinder}}{h} = 1 - \frac{H}{h} = \frac{x}{r}$,

height of the cylinder $= \left(1 - \frac{x}{r}\right) h = H$

$V = \pi x^2 \left(1 - \frac{x}{r}\right) h = \pi x^2 h - \pi \frac{x^3 h}{r}$

differentiating V with respect to x,

$\frac{dV}{dx} = 2\pi x h - \frac{3\pi h x^2}{r}$

this derivative is zero for maximum or minimum

$2\pi x h - \frac{3\pi h x^2}{r} = 0 \quad 2\pi x h = \frac{3\pi h x^2}{r} \quad x = \frac{2r}{3}$.

The second derivative, $\frac{d^2 V}{dx^2} = 2\pi h - \frac{6\pi h}{r} x$.

If $x = \frac{2r}{3}$, $\frac{d^2 V}{dx^2} = 2\pi h - \frac{6\pi h}{r} \frac{2r}{3}$

$= 2\pi h - 4\pi h = -2h\pi$

which is a negative quantity and therefore the volume is a maximum when $x = \frac{2r}{3}$

$V_{max} = \pi \left(\frac{2r}{3}\right)^2 \left(1 - \frac{\frac{2r}{3}}{r}\right) h$

$= \pi \frac{4r^2}{9} \frac{h}{3} = \frac{4}{27} \pi r^2 h$

$V_{max} = \frac{4}{27} \pi r^2 h = \frac{4}{27} \pi 3^2 \left(\frac{3}{4\pi}\right) = 1 \text{ m}^3$.

WORKED EXAMPLE 10

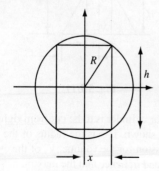

Fig. 5-I/12

A right circular cylinder is to be cut from a solid sphere of radius R as shown in the diagram.

Show that V, the volume of the cylinder, is given by $V = 2\pi x^2 \left(R^2 - x^2\right)^{\frac{1}{2}}$ where x is the radius of the cylinder. Find the maximum volume of the cylinder if x varies and assume that $\frac{d^2 V}{dx^2}$ is negative, that is, no need to find the second derivative.

Solution 10

Let h be the height of the cylinder then by using Pythagoras we obtain

$\frac{1}{2} h = \sqrt{R^2 - x^2}$ or $h = 2\sqrt{R^2 - x^2}$.

The volume of the cylinder

$V = \pi x^2 h = 2\pi x^2 \sqrt{R^2 - x^2}$

$\frac{dV}{dx} = 2\pi (2x) \sqrt{R^2 - x^2} + 2\pi x^2 \frac{1}{2} \left(R^2 - x^2\right)^{-\frac{1}{2}} (-2x)$,

for turning points

$\frac{dV}{dx} = 0, \quad 4\pi x \left(R^2 - x^2\right)^{\frac{1}{2}} = 2\pi x^3 \left(R^2 - x^2\right)^{-\frac{1}{2}}$

$4\pi x \left(R^2 - x^2\right) = 2\pi x^3$

$2\left(R^2 - x^2\right) = x^2 \quad 2R^2 = 3x^2 \quad R = \left(\sqrt{\frac{3}{2}}\right) x$

$V_{max} = 2\pi \frac{2}{3} R^2 \left(R^2 - \frac{2}{3} R^2\right)^{\frac{1}{2}}$

$= \frac{4}{3} \pi R^2 \left(\frac{1}{3} R^2\right)^{\frac{1}{2}}$

$V_{max} = \frac{4\pi}{3\sqrt{3}} R^3$.

WORKED EXAMPLE 11

Determine the turning points of the following functions:-

(i) $y = 2x^2 - 3x - 5$

(ii) $y = -5x^2 - x + 1$

(iii) $y = x^3 + 2x^2 - 3x + 4$

(iv) $y = x^3 - 5x - 7$.

Solution 11

(i) $y = 2x^2 - 3x - 5$, $\dfrac{dy}{dx} = 4x - 3$, $\dfrac{dy}{dx} = 0$

for turning points $4x - 3 = 0$,

$x = \dfrac{3}{4}$ $\dfrac{d^2y}{dx^2} = 4$, $\dfrac{d^2y}{dx^2} > 0$

giving a minimum at $x = \dfrac{3}{4}$,

$y_{min} = 2\left(\dfrac{3}{4}\right)^2 - 3\left(\dfrac{3}{4}\right) - 5 = \dfrac{9}{8} - \dfrac{9}{4} - 5 = -\dfrac{49}{8}$.

The turning point has coordinates $\left(\dfrac{3}{4}, -\dfrac{49}{8}\right)$.

(ii) $y = -5x^2 - x + 1$, $\dfrac{dy}{dx} = -10x - 1$ $\dfrac{dy}{dx} = 0$

for turning points $-10x - 1 = 0$, $x = -\dfrac{1}{10}$

$\dfrac{d^2y}{dx^2} = -10 < 0$ giving a maximum at $x = -\dfrac{1}{10}$

$y_{max} = -5\left(-\dfrac{1}{10}\right)^2 - \left(-\dfrac{1}{10}\right) + 1$

$= -\dfrac{1}{20} + \dfrac{1}{10} + 1 = \dfrac{21}{20}$.

The turning point has coordinates $\left(-\dfrac{1}{10}, \dfrac{21}{20}\right)$.

(iii) $y = x^3 + 2x^2 - 3x + 4$, $\dfrac{dy}{dx} = 3x^2 + 4x - 3$,

$\dfrac{dy}{dx} = 0$ for turning points $3x^2 + 4x - 3 = 0$,

$x = \dfrac{-4 \pm \sqrt{16 + 36}}{6} = \dfrac{-4 \pm 7.21}{6}$

$x = -1.87$ or $x = 0.535$, $\dfrac{d^2y}{dx^2} = 6x + 4$

if $x = -1.87$, $\dfrac{d^2y}{dx^2} = 6(-1.87) + 4 = -7.22$
giving a maximum

if $x = 0.535$, $\dfrac{d^2y}{dx^2} = 6(0.535) + 4 = 7.21$ giving a minimum.

There are two turning points, a maximum at -1.87 and a minimum at 0.535.

$y_{max} = (-1.87)^3 + 2(-1.87)^2 - 3(-1.87) + 4$
$= -6.54 + 6.99 + 5.61 + 4 = 10.06$

$y_{min} = (0.535)^3 + 2(0.535)^2 - 3(0.535) + 4$
$= 3.12$

$(-1.87, 10.06)$ and $(0.535, 3.12)$ are the coordinates of the maximum and minimum respectively.

(iv) $y = x^3 - 5x - 7$, $\dfrac{dy}{dx} = 3x^2 - 5 = 0$ for turning points $3x^2 = 5$

or $x = \pm\sqrt{\dfrac{5}{3}} = \pm 1.29$, $\dfrac{d^2y}{dx^2} = 6x$,

if $x = -1.29$, $\dfrac{d^2y}{dx^2} = -6 \times 1.29 < 0$

if $x = 1.29$, $\dfrac{dy}{dx} = 6 \times 1.29 > 0$

$y_{max} = (-1.29)^3 - 5(-1.29) - 7 = -2.7$

$y_{min} = (1.29)^3 - 5(1.29) - 7 = -11.3$

The coordinates of the turning points are $(-1.29, -2.7)$ and $(1.29, -11.3)$ for maximum and minimum respectively.

Quadratic functions either have a maximum or a minimum cubic functions normally have a maximum and minimum and in most cases the point of inflections may be determined.

Exercises 1

1. Find the coordinates of the turning point on the curve $y = 5e^{2x} + 7e^{-2x}$ and determine the nature of this turning point.

2. For the curve $y = 2\sin^2 x \cos^2 x$, where $0 \leq x \leq 2\pi$, find the x- and y- coordinates of the points at which $\dfrac{dy}{dx} = 0$. Hence, determine the nature of the turning points in the above range.

3. Determine the maximum or minimum values of the following functions:-

   (i) $y = x^2 + 4x + 8$
   (ii) $y = x^2 - 4x + 7$
   (iii) $y = -x^2 - 6x - 7$
   (iv) $y = -x^2 + 6x - 8$.

   Sketch these functions and indicate their turning points.

4. Determine the stationary points on the following curves and distinguish between them:

   (i) $y = x^3 + 2x^2 - 11x - 12$
   (ii) $y = x^3 - x$.

   Sketch the graphs and indicate the turning points.

5. Find the first and second derivatives of the cubic function $y = x^3 - 2x^2 - 15x - 12$. Show that there is a maximum at $x = -\frac{5}{3}$ and a minimum at $x = 3$.

   Sketch the curve.

6. With the aid of the gradient, explain the meaning of the turning point or, stationary point. What is the significance of $\frac{d}{dx}\left(\frac{dy}{dx}\right)$ being

   (a) positive

   (b) negative

   (c) zero.

7. Determine the turning points of the functions:-

   (i) $y = 2x^2 - 3x - 5$

   (ii) $y = -3x^2 - 2x + 1$

   (iii) $y = x^3 + 2x^2 - x + 1$

   (iv) $y = 2x^3 - 4x^2 - 3x - 5$.

8. Determine the nature of the turning points of the functions in question 7.

9. Find the points of inflexion of the curves:-

   (i) $y = x^3 + 2x^2 - x + 1$

   (ii) $y = 2x^3 - 4x^2 - 3x - 5$.

10. Determine the maximum or minimum value of the quadratic functions:-

    (i) $y = 2x^2 + 2x + 2$

    (ii) $y = 2x^2 - 2x + 4$

    (iii) $y = -x^2 - 3x - 4$

    (iv) $y = -x^2 + 5x - 5$.

11. The sum of two numbers is 144, find the numbers such that their product is maximum.

12. An open rectangular tank with a square base is to contain 500 m³ of liquid. If the side of the base is y metres, show that the total surface area S of metal is given by $S = y^2 + \frac{2000}{y}$.

    Find the dimensions of the tank for a minimum surface area.

13. The power P dissipated to the load R by the generator of internal resistance r and e.m.f. E is given by $P = IE - I^2 r$.

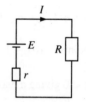

Fig. 5-I/13

Determine the value of the current I for maximum power when $E = 12$ V and $r = 0.9\,\Omega$. Calculate the value of the maximum power for this current.

14. A closed box is to have a volume of 150 m³ and the length of the square base is x m while that of the depth is h m. Determine the dimensions of the box so that the surface area shall be a minimum.

15. An open tank is to be made from rectangular sheet of metal 15 m by 25 m. Equal squares are cut from each of the corners and the sides are bent up form the tank. Determine the maximum volume of the tank.

2

Sketch the Algebraic–Exponential Composite Functions

WORKED EXAMPLE 12

Sketch the curve $y = xe^x$.

Solution 12

$\dfrac{dy}{dx} = e^x + xe^x$

$\dfrac{dy}{dx} = 0, e^x(1+x) = 0 \quad x = -1$

$\qquad e^x = 0 \quad x \to -\infty$ undefined.

$\dfrac{d^2y}{dx^2} = e^x + xe^x + e^x = 2e^x + xe^x$

$\dfrac{d^2y}{dx^2} = 2e^{-1} - e^{-1} = e^{-1} = \dfrac{1}{e} > 0$ minimum

when $x = 0, y = 0$, when $x = -1, y = -e^{-1} = -\dfrac{1}{e}$.

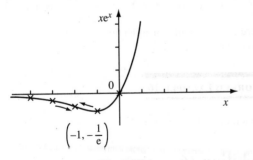

Fig. 5-I/14

The x-axis is an asymptote as $x \to -\infty$

$y = -\dfrac{\infty}{e^\infty} \to 0, \dfrac{d^2y}{dx^2} = 0, x = -2$ has a point of inflection.

WORKED EXAMPLE 13

Sketch the curve $y = xe^{-x}$.

Solution 13

$\dfrac{dy}{dx} = e^{-x} - xe^{-x} \qquad \dfrac{dy}{dx} = 0$ for turning points

$(1-x)e^{-x} = 0, x = 1$

and $x \to \infty, \dfrac{d^2y}{dx^2} = -e^{-x} - e^{-x} + xe^{-x} = -2e^{-x} + xe^{-x} = e^{-x}(-2+x)$.

If $x = 1, \dfrac{d^2y}{dx^2} = -e^{-1} = -\dfrac{1}{e} < 0$ max, $\dfrac{d^2y}{dx^2}$ gives a point of inflection at $x = 2$.

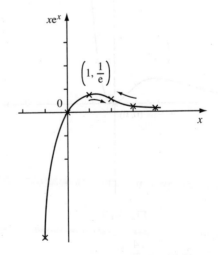

Fig. 5-I/15

WORKED EXAMPLE 14

Sketch the curve $y = x^2 e^x$.

Solution 14

$\dfrac{dy}{dx} = 2xe^x + x^2 e^x \qquad \dfrac{dy}{dx} = 0$ for turning points

$(x^2 + 2x)e^x = 0 \quad e^x \to 0$ when $x \to -\infty$ not defined

$x(x+2) = 0 \quad x = 0, x = -2$

$\dfrac{d^2 y}{dx^2} = 2e^x + 2xe^x + x^2 e^x + 2xe^x$

$= e^x (x^2 + 4x + 2) = 0$

$x = \dfrac{-4 \pm \sqrt{16-8}}{2} = \dfrac{-4 \pm \sqrt{8}}{2}$

$= -2 \pm \sqrt{2}$

there are two points of inflections,

at $x = -2 + \sqrt{2}$, and $x = -2 - \sqrt{2}$ when $x = 0$,
$\dfrac{d^2 y}{dx^2} = 2e^0 > 0$ min.

$x = -2 \qquad \dfrac{d^2 y}{dx^2} = e^{-2}(4 - 8 + 2) = -\dfrac{2}{e^2} < 0$ max.

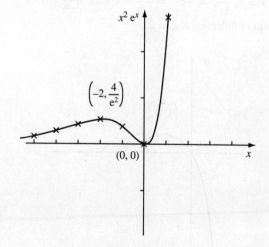

Fig. 5-I/16

when $x = -2$, $y_{max} = \dfrac{4}{e^2}$.

WORKED EXAMPLE 15

Sketch the curve $y = x^2 e^{-x}$.

Solution 15

$\dfrac{dy}{dx} = 2xe^{-x} - x^2 e^{-x} \qquad \dfrac{dy}{dx} = 0$ for turning points

$e^{-x}(2x - x^2) = 0, x = 0, x = 2$ and $x \to \infty$

$\dfrac{d^2 y}{dx^2} = 2e^{-x} + 2x(-e^{-x}) - 2xe^{-x} - x^2(-e^{-x})$

$= (x^2 - 4x + 2)e^{-x}$

when $x = 0$, $\dfrac{d^2 y}{dx^2} = \dfrac{2}{e^0} = 2 > 0$ min.

when $x = 2$, $\dfrac{d^2 y}{dx^2} = (4 - 8 + 2)e^{-2} = -\dfrac{2}{e^2} < 0$ max.

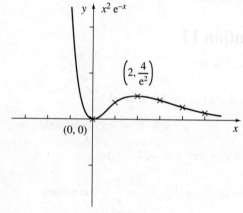

Fig. 5-I/17

$\dfrac{d^2 y}{dx^2} = 0, x^2 - 4x + 2 = 0, x = \dfrac{4 \pm 2\sqrt{2}}{2} = 2 \pm \sqrt{2}$

there are two points of inflection, one at $x = 2 + \sqrt{2}$ and one at $x = 2 - \sqrt{2}$.

WORKED EXAMPLE 16

Sketch the curve $y = x^2 e^{-x^2}$.

Solution 16

$\dfrac{dy}{dx} = 2xe^{-x^2} - x^2 2xe^{-x^2} = 2xe^{-x^2} - 2x^3 e^{-x^2}$

$\dfrac{dy}{dx} = 0$ for turning points $e^{-x^2}(-2x^3 + 2x) = 0, x = 0, x = \pm 1$

$$\frac{d^2y}{dx^2} = 2e^{-x^2} - 4x^2e^{-x^2} - 6x^2e^{-x^2} + 4x^4e^{-x^2}$$

$$= 2e^{-x^2}(2x^4 - 5x^2 + 1) = 0$$

for points of inflections

$2W^2 - 5W + 1 = 0$, where $W = x^2$

$$W = \frac{5 \pm \sqrt{25-8}}{4} = \frac{5 \pm \sqrt{17}}{4}$$

$x^2 = \frac{5+\sqrt{17}}{4}$, $x^2 = \frac{5-\sqrt{17}}{4}$ there are four points of inflections

$x = \pm\sqrt{\frac{5+\sqrt{17}}{2}}$ and $x = \pm\sqrt{\frac{5-\sqrt{17}}{2}}$.

However, when $x = 0$, $\frac{d^2y}{dx^2} = 2 > 0$ min.

when $x = 1$, $\frac{d^2y}{dx^2} = -\frac{4}{e}$ max.

when $x = -1$, $\frac{d^2y}{dx^2} = -\frac{4}{e}$ max.

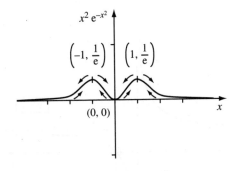

Fig. 5-I/18

Exercises 2

1. A curve is given by the equation $y = xe^x$, show that $x\frac{d^2y}{dx^2} - x\frac{dy}{dx} - y = 0$.

2. A curve is given by the equation $y = xe^{-x}$ show that $\frac{d^2y}{dx^2} + 2\frac{dy}{dx} + y = 0$.

3. A curve is given by the equation $y = e^{-x^2}$ show that $\frac{d^2y}{dx^2} + 2x\frac{dy}{dx} + 2y = 0$.

4. Determine the first and second derivatives of the functions:-
   (i) $y = xe^{-3x}$
   (ii) $y = xe^{2x}$.

5. Differentiate the following algebraic - exponential functions:-
   (i) $(2x-3)^3 e^{-x^2}$
   (ii) $(x+1)^4 e^{x^3}$
   (iii) $x^3 e^{x^3}$.

6. If $y = xe^{kx}$ show that $x\frac{dy}{dx} = y(1+kx)$.

7. If $y = 3xe^{-x}$ show that $\frac{d^2y}{dx^2} + 2\frac{dy}{dx} + y = 0$.

8. If $y = x^2 e^{2x}$ show that $\frac{d^2y}{dx^2} - 4\frac{dy}{dx} + 4y = 2e^{2x}$.

9. Find the first and second derivatives of the following functions:-
   (i) $y = 3xe^{-x}$
   (ii) $y = xe^{-x^2}$
   (iii) $y = \frac{e^{x^2}}{x}$.

10. Find the maximum and minimum values of the functions xe^{-3x^2}. Hence sketch the curve.

3

Algebraic Functions with Asymptotes

WORKED EXAMPLE 17

(i) Sketch the curve $y = \dfrac{1}{1-x}$.

(ii) Sketch the curve $y = \dfrac{x}{1+x}$.

Solution 17

(i) $\dfrac{dy}{dx} = -\dfrac{1(-1)}{(1-x)^2} = \dfrac{1}{(1-x)^2}$, the gradient is always positive

If $x = 0$, $y = 1$, when $x \to 1$, $y \to \infty$ asymptote.

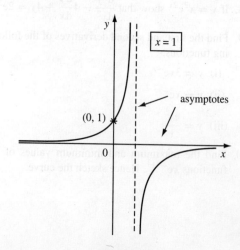

Fig. 5-I/19

$y = \dfrac{1}{x\left(\frac{1}{x} - 1\right)}$, when $x \to \infty$ $y \to 0$

$x \to -\infty$ $y \to 0$ asymptote.

(ii) $y = \dfrac{x}{1+x}$

$\dfrac{dy}{dx} = \dfrac{1(1+x) - x.1}{(1+x)^2} = \dfrac{1}{(1+x)^2}$, the gradient is always positive.

To find the asymptotes

$y = \dfrac{x+1-1}{1+x} = 1 - \dfrac{1}{1+x}$

when $x \to \infty$ $y \to 1$

$x \to -1$ $y \to -\infty$ when $x = 0, y = 0$.

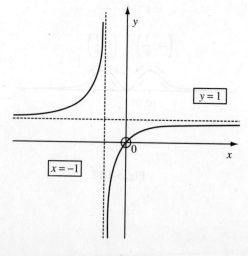

Fig. 5-I/20

WORKED EXAMPLE 18

Sketch the curve $y = \dfrac{x^2}{x-1}$.

Solution 18

$$\begin{array}{r}x+1\\x-1\overline{)x^2}\\\underline{x^2-x}\\x\\\underline{x-1}\\1\end{array}$$

$$\boxed{y = x + 1 + \frac{1}{x-1}}$$

$x \to 1 \quad y \to \infty$

$x \to \infty \quad y \to x + 1$

$$\frac{dy}{dx} = \frac{2x(x-1) - x^2 \cdot 1}{(x-1)^2} = \frac{2x^2 - 2x - x^2}{(x-1)^2}$$

$$= \frac{x^2 - 2x}{(x-1)^2}$$

$\frac{dy}{dx} = 0$ for turning points $\frac{x^2 - 2x}{(x-1)^2} = 0,\; x = 0,\; x = 2$

$$\frac{d^2y}{dx^2} = \frac{(2x-2)(x-1)^2 - (x^2-2x)\,2(x-1)}{(x-1)^4}$$

$$= \frac{(2x-2)(x-1) - 2(x^2-2x)}{(x-1)^3}$$

$$\frac{d^2y}{dx^2} = \frac{2x^2 - 2x - 2x + 2 - 2x^2 + 4x}{(x-1)^3} = \frac{2}{(x-1)^3}$$

when $x = 0,\; \dfrac{d^2y}{dx^2} = -2 < 0$ max.,

$x = 2,\; \dfrac{d^2y}{dx^2} = \dfrac{2}{1} > 0$ min.

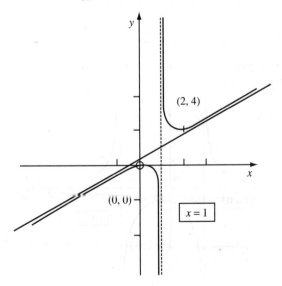

Fig. 5-I/21

when $x = 0,\; y = 0$

$x = 2,\; y = \dfrac{4}{2-1} = 4$

when $x = -1,\; y = \dfrac{1}{-2} = -\dfrac{1}{2}$

$x = -5,\; y = \dfrac{25}{-6}$

Worked Example 19

Sketch the curve $y = \dfrac{x}{(x-1)(x+2)}$.

Solution 19

Express the function in partial fractions

$$y = \frac{x}{(x-1)(x+2)} \equiv \frac{A}{x-1} + \frac{B}{x+2}$$

$x \equiv A(x+2) + B(x-1)$.

If $x = 1,\; A = \dfrac{1}{3}$; and if $x = -2,\; B = \dfrac{2}{3}$

$$y = \frac{\frac{1}{3}}{x-1} + \frac{\frac{2}{3}}{x+2}$$

$$= \left(\frac{1}{3}\right)(x-1)^{-1} + \left(\frac{2}{3}\right)(x+2)^{-1}$$

differentiating with respect to x

$$\frac{dy}{dx} = -\frac{1}{3}(x-1)^{-2} - \frac{2}{3}(x+2)^{-2}$$

$$= -\frac{1}{3(x-1)^2} - \frac{2}{3(x+2)^2}$$

the gradient is always negative. There are no turning points. The asymptotes are found as follows:- when $x \to 1$ or $x \to -2,\; y \to \infty$, when $x \to \infty,\; y \to 0$.

cross multiplying the function

$y(x-1)(x+2) = x \Rightarrow y(x^2 + x - 2) - x = 0$

$x^2 y + xy - 2y - x = 0 \Rightarrow x^2 y + x(y-1) - 2y = 0$

a quadratic equation in x. For x to be real, the discriminant $D = b^2 - 4ac \geq 0,\; (y-1)^2 + 8y^2 \geq 0$ therefore y exists for all negative or positive values.

16 — GCE A level

When $x = 0$, $y = 0$ when $x = -1$, $y = \dfrac{1}{2}$

$x = \dfrac{1}{2}$, $y = \dfrac{\frac{1}{2}}{\left(-\frac{1}{2}\right)\left(\frac{5}{2}\right)} = -\dfrac{2}{5}$.

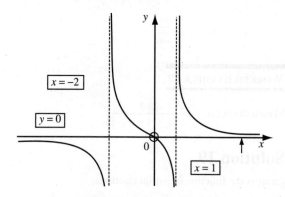

Fig. 5-I/22

WORKED EXAMPLE 20

Sketch the curve $y = \dfrac{1}{(x+2)(x-3)}$.

Solution 20

Expressing in partial fractions

$$\dfrac{1}{(x+2)(x-3)} \equiv \dfrac{A}{x+2} + \dfrac{B}{x-3}$$

$$1 \equiv A(x-3) + B(x+2)$$

if $x = 3$, $B = \dfrac{1}{5}$, if $x = -2$, $A = -\dfrac{1}{5}$

$$y = \dfrac{-\frac{1}{5}}{x+2} + \dfrac{\frac{1}{5}}{x-3}$$

$$y = -\dfrac{1}{5}(x+2)^{-1} + \dfrac{1}{5}(x-3)^{-1}$$

$$\dfrac{dy}{dx} = +\dfrac{1}{5}(x+2)^{-2} - \dfrac{1}{5}(x-3)^{-2}$$

$$\dfrac{d^2y}{dx^2} = -\dfrac{2}{5}(x+2)^{-3} + \dfrac{2}{5}(x-3)^{-3}.$$

For turning points $\dfrac{dy}{dx} = 0$

$$\dfrac{1}{5}(x+2)^{-2} - \dfrac{1}{5}(x-3)^{-2} = 0$$

$$(x+2)^{-2} = (x-3)^{-2}$$

$$\dfrac{(x-3)^2}{(x+2)^2} = 1$$

square rooting both sides, we have $\dfrac{x-3}{x+2} = \pm 1$

$\dfrac{x-3}{x+2} = 1$, $x - 3 = x + 2$ there is no solution

$\dfrac{x-3}{x+2} = -1$, $x - 3 = -x - 2$, $2x = 1$, $x = \dfrac{1}{2}$.

$$\dfrac{d^2y}{dx^2} = -\dfrac{2}{5(x+2)^3} + \dfrac{2}{5(x-3)^3}$$

$$= -\dfrac{2}{5\left(\frac{1}{2}+2\right)^3} + \dfrac{2}{5\left(\frac{1}{2}-3\right)^3}$$

$$= -\dfrac{2 \times 8}{5 \times 125} - \dfrac{2 \times 8}{5 \times 125} = -\dfrac{32}{625}$$

$\dfrac{d^2y}{dx^2} < 0$ giving a maximum

$$y_{max} = \dfrac{-\frac{1}{5}}{\frac{1}{2}+2} + \dfrac{\frac{1}{5}}{\frac{1}{2}-3} = \dfrac{-\frac{1}{5}}{\frac{5}{2}} - \dfrac{\frac{1}{5}}{\frac{5}{2}}$$

$$y_{max} = -\dfrac{4}{25}$$

Asymptotes $x = -2$ or $x = 3$ when $y \to \infty$, $x \to \infty$ when $y \to 0$.

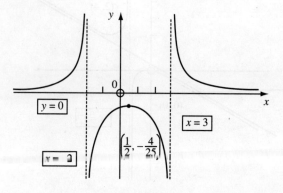

Fig. 5-I/23

$\dfrac{dy}{dx} = \dfrac{1}{5(x+2)^2} - \dfrac{1}{5(x-3)^2}$, if $x < -2$, that is, when $x = -3$

$\dfrac{dy}{dx} = \dfrac{1}{5} - \dfrac{1}{5 \times 36} = \dfrac{7}{36} > 0$ the gradient is positive

if $x > 3$, that is, when $x = 4$,

$\dfrac{dy}{dx} = \dfrac{1}{5 \times 36} - \dfrac{1}{5} = -\dfrac{7}{36} < 0$, the gradient is negative.

Exercises 3

1. Express $\dfrac{x}{(2x+1)(x-3)}$ in partial fractions and hence find the value of

    $\dfrac{d^2 y}{dx^2}$ when $x = -2$, to three decimal places.

2. Express $\dfrac{x^2 + 3x - 1}{(1+x)^2(1-x)}$ in partial fractions.

    Evaluate $\dfrac{d^2 y}{dx^2}$ when $x = 2$, to two decimal places.

3. Sketch the function $y = \dfrac{1}{x^2 - 1}$.

4. Sketch the function $y = \dfrac{x}{x - 1}$.

5. Evaluate $\dfrac{d^3 y}{dx^3}$ when $x = 0$, if $y = \dfrac{x^2}{(x^2 - 1)(x + 3)}$.

6. Express $y = \dfrac{x^2 - 2x - 4}{(x^2 - 4)(x + 1)}$ in partial fractions and hence find the first and second derivatives when $x = 0$.

7. Sketch the curve $\dfrac{x}{1 - x^2}$. Are there any turning points?

8. Find the turning points of $y = \dfrac{x^2 + 1}{5x + 2}$ and hence sketch the curve.

4

Simple Transformations

Transformation from f(x) to f(kx).

WORKED EXAMPLE 21

Sketch the quadratic function $y = f(x) = x^2 + 4x + 4$.

Solution 21

This function has a minimum, since the coefficient of x^2 is positive, this minimum occurs at

$$x = -\frac{b}{2a} = -\frac{4}{2 \times 1} = -2,$$ also $f(0) = 4$ and

$f(x) = 0$ when $x = -2, f(-2) = (-2)^2 - 8 + 4 = 0$.

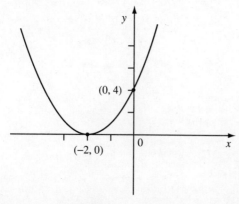

Fig. 5-I/24 $f(x) = (x+2)^2$

WORKED EXAMPLE 22

If x is replaced by $2x$ or $\frac{1}{2}x$, in the previous example, we then have $f(2x) = (2x + 2)^2 = 4(x + 1)^2$ and $f\left(\frac{1}{2}x\right) = \left(\frac{1}{2}x + 2\right)^2$. Sketch these graphs.

Solution 22

$f(2x) = 4(x + 1)^2 = 4x^2 + 8x + 4$, this function has a minimum and it occurs at $x = -\frac{b}{2a} = -\frac{8}{2\times 4} = -1$, $f(0) = 4$ again, and $f(2x) = 0$ when $x = -1$.

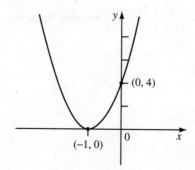

Fig. 5-I/25

For each point (x, y) on the graph of $y = f(x)$, there exists a point $\left(\frac{x_1}{k}, y_1\right)$ on the graph of $y = f(kx)$.

The graph of $y = f(kx)$ can be obtained by sketching $y = f(x)$ parallel to the x-axis by a scale factor $\frac{1}{k}$.

So that $f(x) = (x + 2)^2$ can be transformed to $f(2x) = (2x + 2)^2 = 4(x + 1)^2$ by stretching $y = f(x)$ parallel to the x-axis by a scale factor $\frac{1}{2}$.

$$f\left(\frac{1}{2}x\right) = \left(\frac{1}{2}x + 2\right)^2$$

$$= \frac{1}{4}x^2 + 2x + 4.$$

The minimum occurs when $x = -\frac{b}{2a}$

$$= -\frac{2}{2\left(\frac{1}{4}\right)} = -4$$

$f(0) = 4$ again, and $f\left(\tfrac{1}{2}x\right) = 0$ when $x = -4$. The scaled factor now is $\tfrac{1}{\frac{1}{2}} = 2$, the graph is stretched by a scale factor 2.

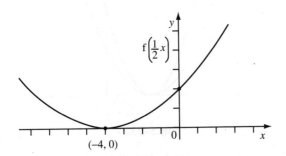

Fig. 5-I/26

Combining the three graphs together, we can see, that $f(x)$ contracts to $f(2x)$ and stretches to $f\left(\tfrac{1}{2}x\right)$.

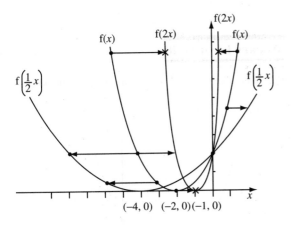

Fig. 5-I/27

Transformation $f(x)$ to $f(x) + k$.

The graph $f(x)$ is transformed to the graph $f(x) + k$ by translating k units parallel to the x-axis. Thus for every point (x, y) on $y = f(x)$, there exists a point $(x_1, y_1 + k)$ on $y = f(x) + k$.

WORKED EXAMPLE 23

$y = f(x) = (x+2)^2$ is transformed to the graph $y = f(x) + 2 = (x+2)^2 + 2$ by shifting the graph vertically by 2.

Solution 23

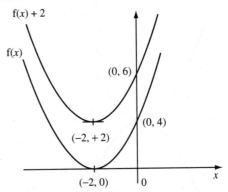

Fig. 5-I/28

Transformation of $y = f(x)$ to $y = f(x - k)$.

Thus for every point (x_1, y_1) on $y = f(x)$, there exists a point $(x_1 + k, y_1)$ on $y = f(x - k)$.

WORKED EXAMPLE 24

Transform $y = f(x) = (x+2)^2$ to $y = f(x+2) = (x+2+2)^2 = (x+4)^2$.

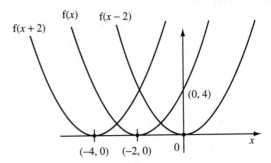

Fig. 5-I/29

Solution 24

The graph is translated by 2 units parallel to the y-axis.

WORKED EXAMPLE 25

Transformation of $y = f(x)$ to $y = -f(x)$.

Solution 25

For each point (x_1, y_1) on the graph of $y = f(x)$ there exists a point $(x_1, -y_1)$ on $y = -f(x)$. In this case, the graph is reflected in the x-axis.

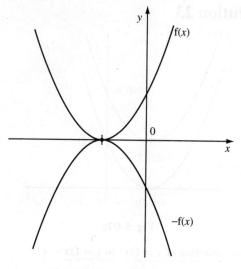

Fig. 5-I/30

WORKED EXAMPLE 26

Transform $y = f(x)$ to $y = f(-x)$.

Solution 26

For each point (x_1, y_1) on the graph $y = f(x)$, there exists a point $(-x_1, y_1)$ oy $y = f(-x)$. Thus the graph of $y = f(x)$ can be obtained by reflecting $y = f(x)$ on the y-axis.

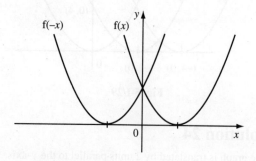

Fig. 5-I/31

WORKED EXAMPLE 27

Transform $y = f(x)$ to $y = kf(x)$.

Solution 27

For each point (x_1, y_1) on the graph of $y = f(x)$ there exists a point (x_1, ky_1) on the graph of $y = kf(x)$.

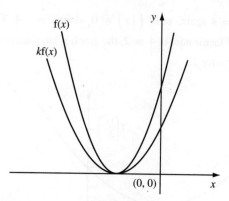

Fig. 5-I/32

The graph $f(x)$ is stretched parallel to the y-axis by a scale factor k.

Graph of Reciprocal

Use the graph $y = f(x)$ to obtain the sketch of the reciprocal graph, $y = \dfrac{1}{f(x)}$.

WORKED EXAMPLE 28

If $y = f(x) = (x+2)^2 = x^2 + 4x + 4$, find the sketch of $\dfrac{1}{f(x)}$.

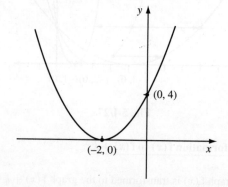

Fig. 5-I/33

Solution 28

$f(x) > 0$ in the range $-2 < x < \infty$ and $-\infty < x < -2$, therefore $\dfrac{1}{f(x)} > 0$.

$f(x) = 0$ when $x = -2$, therefore $\dfrac{1}{f(x)}$ is not defined for this value of x and $x = -2$ is a vertical asymptote of

$y = \frac{1}{f(x)}$ since $f(x)$ cuts the y-axis at the point $(0, 4)$, then $\frac{1}{f(x)}$ will cut the y-axis at the point $\left(0, \frac{1}{4}\right)$.

As $f(x) \to \pm\infty$, then $\frac{1}{f(x)} \to 0$.

$\frac{1}{f(x)} = \frac{1}{(x+2)^2}$ for the domains $-\infty < x < -2$,

$-2 < x < \infty \quad \frac{1}{f(x)} > 0$.

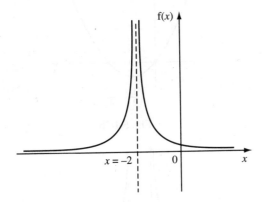

Fig. 5-I/34

WORKED EXAMPLE 29

Sketch the graph $f(x) = (x-1)(x+3)$ and hence sketch the graph $\frac{1}{f(x)}$.

Solution 29

$f(x) = (x-1)(x+3) = x^2 + 2x - 3$.

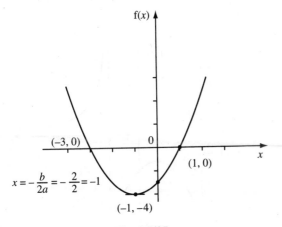

Fig. 5-I/35

This graph has a minimum at $x = -\frac{b}{2a} = -\frac{2}{2} = -1$, $f(-1) = -4$.

$\frac{1}{f(x)} = \frac{1}{(x-1)(x+3)}$

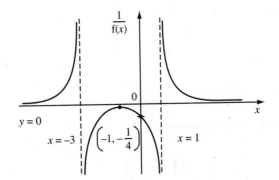

Fig. 5-I/36

There are two asymptotes, at $x = 1$ and at $x = -3$.

There is a maximum at $\left(-1, -\frac{1}{4}\right)$. It crosses the y-axis at $\left(0, -\frac{1}{3}\right)$.

$\frac{1}{f(x)} < 0$ when $-3 < x < 1$

$\frac{1}{f(x)} > 0$ when $-\infty < x < -3$ when $1 < x < +\infty$.

WORKED EXAMPLE 30

Prove that, for all real values of x, $\frac{1}{(x-4)(x+3)} \geq 0$

and $\frac{1}{(x-4)(x+3)} \leq -\frac{4}{49}$.

Sketch the curve $f(x) = \frac{1}{x^2 - x - 12}$.

Solution 30

Let $y = \frac{1}{(x-4)(x+3)}$, $y(x-4)(x+3) = 1$

$y(x^2 - x - 12) = 1$, $yx^2 - yx - 12y - 1 = 0$.

For all real values of x, the discriminant is greater than or equal to zero.

$(-y)^2 - 4y(-12y - 1) \geq 0$, $y^2 + 48y^2 + 4y \geq 0$

$49y^2 + 4y \geq 0 \quad y(49y + 4) \geq 0$.

22 — GCE A level

	$y > 0$	$-\frac{4}{49} < y < 0$	$y < -\frac{4}{49}$
y	+	−	−
$49y + 4$	+	+	−
$y(49y + 4)$	+	−	+

Therefore, $y > 0$ and $y < -\dfrac{4}{49}$. $y \geq 0$, or

$$\frac{1}{(x-4)(x+3)} \geq 0 \quad \text{and} \quad y \leq \frac{4}{49}$$

or $\dfrac{1}{(x-4)(x+3)} \leq -\dfrac{4}{49}$.

There are two asymptotes, $x = -3$ and $x = 4$.

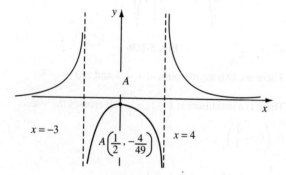

Fig. 5-I/37

The quadratic function $Q(x) = (x-4)(x+3) = x^2 - x - 12$ has a minimum at

$$x = -\frac{b}{2a} = \frac{1}{2},$$

$$f\left(\frac{1}{2}\right) = \left(-\frac{7}{2}\right)\left(\frac{7}{2}\right) = -\frac{49}{4}.$$

The reciprocal of $Q(x)$ is $f(x)$ which has a maximum point at $x = \frac{1}{2}$ and hence the turning point is $\left(\frac{1}{2}, -\frac{4}{49}\right)$.

WORKED EXAMPLE 31

Sketch the following graphs and their reciprocals.

(i) $f(x) = x(x-1)$

(ii) $f(x) = (x-1)(x+2)$

(iii) $f(x) = (x+1)(x-2)(x+3)$

(iv) $f(x) = (x+2)(x-3)^2$

(v) $f(x) = (x-1)^2$.

Solution 31

(i) $f(x) = x(x-1) = x^2 - x$, this has a minimum which occurs at the point,

$$x = -\frac{b}{2a} = \frac{1}{2}, f\left(\frac{1}{2}\right) = \frac{1}{4} - \frac{1}{2} = -\frac{1}{4}, f(x) = 0$$

when $x = 1$ and $x = 0$.

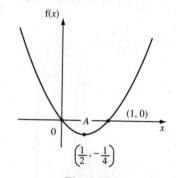

Fig. 5-I/38

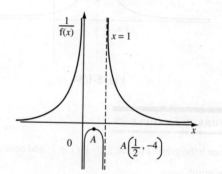

Fig. 5-I/39

(ii) $f(x) = (x-1)(x+2) = x^2 + x - 2$, this has a minimum which occurs at the point, $x = -\frac{1}{2}$, $f\left(-\frac{1}{2}\right) = -\frac{9}{4}$, $f(0) = -2$, $f(x) = 0$ at $x = 1$ and $x = -2$,

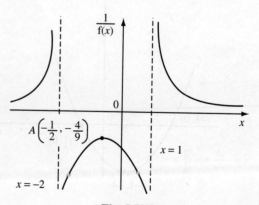

Fig. 5-I/40

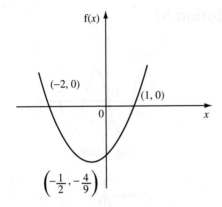

Fig. 5-I/41

(iii) $f(x) = (x+1)(x-2)(x+3)$

$= (x^2 - x - 2)(x+3)$

$= x^3 - x^2 - 2x + 3x^2 - 3x - 6$

$= x^3 + 2x^2 - 5x - 6$

$f'(x) = 3x^2 + 4x - 5 = 0$ for turning points,

$x = \dfrac{-4 \pm \sqrt{16 + 60}}{6} = \dfrac{-4 \pm 8.72}{6}, x = 0.786$ or $x = -2.12$,

$f''(x) = 6x + 4, f''(0.786) = 6 \times 0.786 + 4 > 0$ minimum

$f''(-2.12) = 6 \times (-2.12) + 4$

$= -12.72 + 4 = -8.72 < 0$ maximum

$f(-2.12) = (-1.12)(-4.12)(0.88) = 4.06$

$f(0.786) = 1.786 \times (-1.214) \times 3.786 = -8.21$

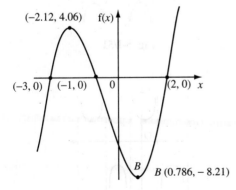

Fig.5-I/42

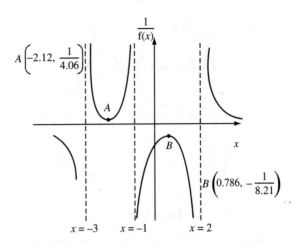

Fig. 5-I/43

(iv) $f(x) = (x+2)(x-3)^2$

$= (x+2)(x^2 - 6x + 9)$

$= x^3 - 6x^2 + 9x + 2x^2 - 12x + 18$

$= x^3 - 4x^2 - 3x + 18$.

$f'(x) = 3x^2 - 8x - 3 = 0$ for turning points,

$x = \dfrac{8 \pm \sqrt{64 + 36}}{6} = \dfrac{8 \pm 10}{6}$

$x = 3$ or $x = -\dfrac{1}{3}$.

$f''(x) = 6x - 8, f''(3) = 18 - 8 = 10$, minimum,

$f''\left(-\dfrac{1}{3}\right) < 0$, maximum. $f(3) = 0$,

$f\left(-\dfrac{1}{3}\right) = \left(\dfrac{5}{3}\right)\left(-\dfrac{10}{3}\right)^2 = \dfrac{500}{27} = 18.52$

$f(x) = 0; x = -2, x = 3$.

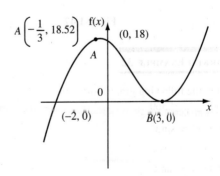

Fig.5-I/44

24 — GCE A level

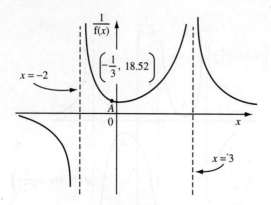

Fig. 5-I/45

(v) $f(x) = (x-1)^2 = x^2 - 2x + 1$, the minimum occurs at $x = 1, f(1) = 0$

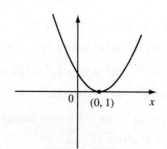

Fig. 5-I/46

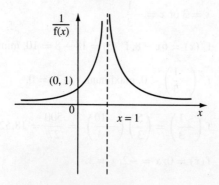

Fig. 5-I/47

WORKED EXAMPLE 32

Sketch the following graphs for range $-360° \leq x \leq 360°$

(i) $f(x) = \sin x$

(ii) $f(x) = \cos x$

(iii) $f(x) = \tan x$

and hence sketch the reciprocal graphs in the same range.

Solution 32

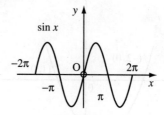

Fig. 5-I/48

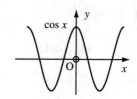

Fig. 5-I/49

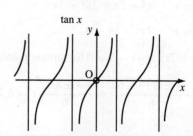

Fig. 5-I/50

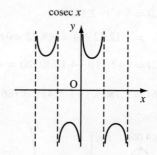

Fig. 5-I/51

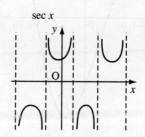

Fig. 5-I/52

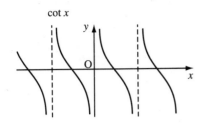

Fig. 5-I/53

Exercises 4

1. If $f(x) = (x-1)^2$, sketch $f(3x)$ and $f\left(\frac{1}{2}x\right)$.

2. If $f(x) = (x^2 + 4x + 7)$, sketch $f(2x)$ and $f\left(\frac{1}{3}x\right)$.

3. If $f(x) = x^2 + 3x + 4$, sketch $f(2x)$ and $f\left(\frac{1}{2}x\right)$.

4. If $f(x) = x^2 + 2x + 4$, sketch $f(x) + 2$ and $f(x) - 3$.

5. If $f(x) = -x^2 + 4x + 1$, sketch $f(x) + 1$ and $f(x) - 1$.

6. If $f(x) = x^2 + 2x + 5$, sketch $f(x-2)$ and $f(x+2)$.

7. If $f(x) = x^2 - 6x + 9$, sketch $-f(x)$ and $f(-x)$.

8. If $f(x) = -x^2 + 2x + 1$, sketch $2f(x)$ and $-3f(x)$.

9. If $f(x) = (x+3)(x-5)$, sketch $\frac{1}{f(x)}$.

10. If $f(x) = (x-4)(x+3)$, sketch $\frac{1}{f(x)}$.

11. If $f(x) = (x-1)(x-2)(x-3)$, sketch $\frac{1}{f(x)}$.

12. If $f(x) = \operatorname{cosec} x$ sketch $\frac{1}{f(x)}, 0 \le x \le 2\pi$.

13. If $f(x) = \sec x$, sketch $\frac{1}{f(x)}, 0 \le x \le 2\pi$.

14. If $f(x) = \cot x$, sketch $\frac{1}{f(x)}, 0 \le x \le 2\pi$.

15. The diagram shows the graph of $y = f(x)$.

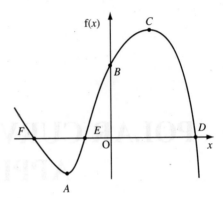

Fig. 5-I/54

The points **A, B, C, D, E, F** have coordinates $(-5, -10)$ $(0, 20)$, $(5, 30)$, $(10, 0)$, $(-3, 0)$ and $(-11, 0)$ respectively. Sketch separately the graph of:-

(i) $y = 2f(x)$

(ii) $y = -f(x)$

(iii) $y = f(-x)$

(iv) $y = f(x-1)$

(v) $y = \frac{1}{f(x)}$

(vi) $y = f(2x)$

(vii) $y = \frac{1}{f(-x)}$.

POLAR CURVE SKETCHING AND APPLICATIONS

Polar Coordinates System

The cartesian coordinates of a point P are x and y and the point P is represented by $P(x, y)$. The cartesian axes are intersected perpendicularly at a point O, called the origin. The point P is plotted as shown in Fig. 5-I/55, x units horizontally and y units vertically.

Let $OP = r$ = the radius vector
= the magnitude of OP and
Θ = the vectorial angle
= the direction of OP with the ox axis.
(positive)

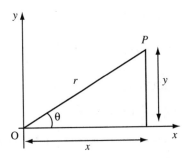

Fig. 5-I/55

In the polar coordinate system, a fixed point is called the pole and a fixed line is called the <u>initial line</u> (a horizontal line drawn to the right of the pole) are used, as shown in Fig. 5-I/56.

A line from the pole O is drawn at an angle Θ positive to the initial line in an anticlockwise direction; a clockwise direction corresponds to a negative angle. This is illustrated in Fig. 5-I/57. The magnitude of $OP = r$ and its direction to the initial line is Θ and therefore the polar coordinates of P are r, Θ and the point $P(r, \Theta)$. Therefore, the position of a point P in a plane is fixed if the distance OP, r, and the angle Θ are known.

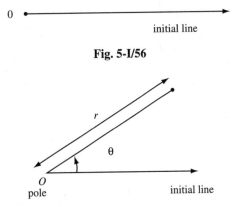

Fig. 5-I/56

Fig. 5-I/57

Sign Conventions

If r is positive in the direction $OP(r > 0)$, then $OP'(-r, \Theta)$ in the direction Θ from the initial line. Fig. 5-I/58 illustrates the sign convention $P(r, \Theta)$ direction $\overrightarrow{OP}$. A negative value of r is in the direction $\overrightarrow{PO}$ produced.

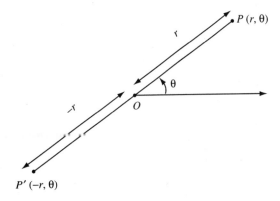

Fig. 5-I/58

Exercises 5

1. Plot the following polar coordinates:-

   (i) $A\left(1, \dfrac{\pi}{6}\right)$

   (ii) $B\left(-1, \dfrac{\pi}{6}\right)$

   (iii) $C\left(2, \dfrac{\pi}{3}\right)$

   (iv) $D\left(-2, \dfrac{\pi}{3}\right)$

   (v) $E\left(-3, \dfrac{\pi}{4}\right)$

   (vi) $F\left(2, \dfrac{3\pi}{4}\right)$

   (vii) $G\left(\sqrt{2}, \dfrac{\pi}{8}\right)$

   (viii) $H\left(\sqrt{3}, 0^c\right)$

   (ix) $I\left(a, \dfrac{\pi}{6}\right)$ where $a > 0$

   (x) $J\left(a, \dfrac{\pi}{6}\right)$ where $a < 0$

   (xi) $K\left(5, \pi^c\right)$

   (xii) $L\left(5, 2\pi\right)$

   (xiii) $M\left(5, \dfrac{\pi}{2}\right)$

   (xiv) $N\left(-5, \dfrac{\pi}{2}\right)$

2. Plot the following pairs of points and calculate the distances PQ in each case.

   (i) $P\left(1, \dfrac{\pi}{6}\right), Q\left(-1, \dfrac{\pi}{6}\right)$

   (ii) $P\left(2, \dfrac{\pi}{3}\right), Q\left(-3, \dfrac{\pi}{4}\right)$

   (iii) $P\left(\sqrt{3}, 0^c\right), Q\left(-\sqrt{3}, 0^c\right)$

   (iv) $P\left(3, \dfrac{\pi}{2}\right), Q\left(-3, -\dfrac{\pi}{2}\right)$

   (v) $P\left(4, \tan^{-1}\dfrac{1}{2}\right) Q\left(-4, \tan^{-1}\dfrac{1}{2}\right)$

   (vi) $P\left(0, \dfrac{\pi}{4}\right) Q\left(3, \dfrac{\pi}{4}\right)$.

3. Find the squares of the distances between the following pairs of points:-

   (i) $P\left(3, 0\right), Q\left(2, \dfrac{\pi}{8}\right)$

   (ii) $P\left(-3, \dfrac{\pi}{3}\right), Q\left(-1, \dfrac{2\pi}{3}\right)$

   (iii) $P\left(-2, \dfrac{7\pi}{4}\right), Q\left(2, \dfrac{\pi}{3}\right)$

   (iv) $P\left(1, 0\right), Q\left(1, \dfrac{\pi}{2}\right)$

   (v) $P\left(-2, \pi\right), Q\left(-2, \dfrac{3\pi}{2}\right)$.

4. A triangle ABC has the following polar coordinates:-
   $A(1, 0), B\left(2, \dfrac{\pi}{2}\right)$ and $C(3, \pi)$.
   Determine the angles A, B and C of the triangle.

6

Relationship between Polar and Cartesian Coordinates

The cartesian coordinates may be expressed in terms of r and Θ. (x, y), cartesian coordinates, and (r, Θ), polar coordinates.

Referring to Fig. 5-I/55, we have $x = r \cos \Theta$
$y = r \sin \Theta$

$r = \sqrt{x^2 + y^2} \qquad \Theta = \tan^{-1} \frac{y}{x}.$

$(r, \Theta) \equiv \left(\sqrt{x^2 + y^2}, \tan^{-1} \frac{y}{x} \right).$

Therefore, a curve given in cartesian coordinates can be expressed in polar coordinates and vice-versa.

Exercises 6

1. Find the polar equations of the following linear cartesian equations:-

    (i) $\dfrac{x}{a} + \dfrac{y}{b} = 1$

    (ii) $y = mx + c$

    (iii) $ax + by + c = 0$

    (iv) $\dfrac{x}{1} + \dfrac{y}{2} = 1$

    (v) $\dfrac{x}{-2} + \dfrac{y}{3} = 1$

    (vi) $y = -3x - 5$

    (vii) $x + 2y - 3 = 0$

    (viii) $y = -3x$

    (ix) $y = x$

    (x) $x = 5$

    (xi) $x = -3$

    (xii) $y = 2$

    (xiii) $y = -5$

    (xiv) $y = 0$

    (xv) $x = 0$

    (xvi) $y = x + 1$

    (xvii) $y = -x - 2$

    (xviii) $y = 2x$

    (xix) $x = -10$

    (xx) $x + y = 1$.

2. Find the polar equations of the following cartesian curves:-

    (i) $x^2 + y^2 = 1$

    (ii) $x^2 + y^2 = 5^2$

    (iii) $\dfrac{x^2}{3^2} + \dfrac{x^2}{4^2} = 1$

    (iv) $-\dfrac{y^2}{5^2} + \dfrac{x^2}{4^2} = 1$

    (v) $xy = 5$

    (vi) $y^2 = 4x$

    (vii) $x^2 = -4y$

    (viii) $y^2 = 4(x - 5)$

    (ix) $x^2 = y - 1$

(x) $y = y^2 - 3x - 1$

(xi) $xy = c^2$

(xii) $\dfrac{x^2}{a^2} + \dfrac{y^2}{b^2} = 1$

(xiii) $\dfrac{x^2}{a^2} - \dfrac{y^2}{a^2} = 1$

(xiv) $(x-3)^2 + (y+5)^2 = 5^2$

(xv) $x^2 + y^2 - x - y - 1 = 0$

(xvi) $y = \dfrac{1}{x}$

(xvii) $xy + c^2 = 0$.

3. Find the cartesian equations of the following polar equations:-

(i) $r = 2\cos 2\Theta$

(ii) $r = 3\sin 2\Theta$

(iii) $r = 5\cos 3\Theta$

(iv) $r = 4\sin 3\Theta$

(v) $r = \sqrt{2} + \cos\Theta$

(vi) $r = \sqrt{3} - \sin\Theta$

(vii) $r = 1 + \cos\Theta$

(viii) $r = 1 + \sin\Theta$

(ix) $r = 1 - \cos\Theta$

(x) $r = 1 - \sin\Theta$

(xi) $r^2 = 2\sin 2\Theta$

(xii) $r^2 = 3\cos 2\Theta$

(xiii) $r = 3 + \cos\Theta$

(xiv) $r = 1 + 2\cos\Theta$

(xv) $r = 5\Theta$

(xvi) $r = -3\Theta$

(xvii) $\Theta = \dfrac{\pi}{3}$

(xviii) $\Theta = -\dfrac{2\pi}{3}$

(xix) $r = 3\sec\left(\Theta - \dfrac{\pi}{2}\right)$

(xx) $r = 5\sec\Theta$

(xxi) $r = \sec\left(\dfrac{\pi}{2} - \Theta\right)$

(xxii) $r = 3\csc\Theta$

(xxiii) $r = 3$

(xxiv) $r = 3\sin\Theta$

(xxv) $r = 5\cos\Theta$

(xxvi) $r^2 = a^2\cos 2\Theta$

(xxvii) $r^2 = a^2\sin 2\Theta$

(xxviii) $r = \cos^2 2\Theta$

(xxix) $r = \sin^2 2\Theta$

(xxx) $r = 3\sec(\Theta - \alpha)$ where $0 \leq \alpha \leq \dfrac{\pi}{2}$

(xxxi) $r = 2\cos(\Theta + \pi)$.

4. Sketch the polar equations shown in exercise 3 above, where r and Θ are polar coordinates.

7

Half-Lines or Part-Lines

The polar equations of half-lines are shown in Fig. 5-I/59. The polar equation of the initial line is $\Theta = 0^c$ in radians, or $\Theta = 0°$ in degrees.

NOTE that the lengths of these half-lines are indefinite $r \geq 0$.

Any line drawn from the pole O at an angle Θ from the initial line is called, the half-line. The initial line is the half-line with polar equation $\Theta = 0°$.
Θ is independent of r.

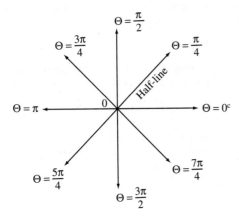

Fig. 5-I/59

Exercises 7

1. Sketch the following half-lines.

   (i) $\Theta = 0^c, \Theta = \dfrac{\pi}{4}, \Theta = \dfrac{\pi}{2}, \Theta = \pi, \Theta = \dfrac{3\pi}{2}$

   (ii) $\Theta = 30°, \Theta = 60°, \Theta = 120°$.

2. Explain the meaning of a half-line or a part-line.

3. The polar equation of a line is given as $\Theta = \alpha$ where α is an angle expressed in degrees or radians.

   Sketch this line if α is an acute angle. How long is this line?

   Note that Θ is independent of α what is a half-line or a part-line?

4. Give the polar equation of the half-line or the part-line.

   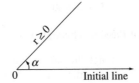

   Write down the limits of half-line.

5. Sketch the following half-lines:

   (i) $\Theta = \dfrac{-3\pi}{2}$, (ii) $\Theta = \dfrac{-\pi}{2}$, (iii) $\Theta = -2\pi$.

8

The Polar Equations of Straight Lines

(a) A Line Passing Through The Pole

A straight line is drawn from the pole, o, as shown in Fig. 5-I/60, in an anti-clockwise direction of an angle α and is represented by

$\boxed{\Theta = \alpha}$ which is called the half-line equation.

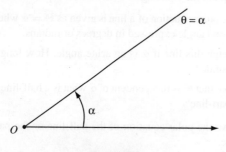

Fig. 5-I/60

(b) A Line Not Passing Through The Pole

A line not passing through the pole may be now considered, making an angle ϕ with the initial line as shown in Fig. 5-I/61.

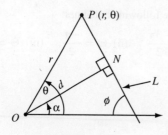

Fig. 5-I/61

Drawing a perpendicular line from the pole, O, to the line L as shown, making an angle α with the initial line, let N be the foot of this perpendicular, and $ON = d$.

A point P is chosen on the line with polar coordinates (r, Θ).

From the triangle ONP, $\dfrac{ON}{OP} = \cos(\Theta - \alpha)$

$OP = r = ON \sec(\Theta - \alpha)$.

Therefore, $\boxed{r = d \sec(\Theta - \alpha)}$ is the polar equation of the line L.

In cartesian coordinate system, a vertical line is represented by the equation $x = k$ where k is a constant.

In polar coordinate system this vertical line can be represented by the equation

$r \cos \Theta = k$, then $\boxed{r = k \sec \Theta}$ where $d = k$ and $\alpha = 0$.

Similarly, a horizontal line may be represented as $r \sin \Theta = k$, or $r = k \operatorname{cosec} \Theta$ or $r = k \sec(\Theta - \frac{\pi}{2})$

$\boxed{r = k \sec\left(\Theta - \dfrac{\pi}{2}\right)}$ where $d = k$ and $\alpha = \frac{\pi}{2}$.

Fig. 5-I/62 and Fig. 5-I/63 represent a vertical line perpendicular to the initial line and a horizontal line to the initial line respectively.

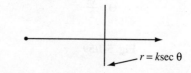

Fig. 5-I/62

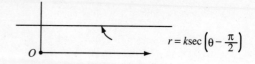

Fig. 5-I/63

Exercises 8

1. In cartesian coordinate system, a line is given as $x = 3$. Write down the polar equation of this line.

2. In cartesian coordinate system, a line is given as $x = 0$. Write down the polar equation of this line.

3. In cartesian coordinate system, a line is given as $y = 2$. Write down the polar equation of this line.

4. The polar equation of a line is given as $r = 3 \sec\left(\Theta - \frac{\pi}{2}\right)$. Determine the corresponding cartesian equation, and show that $3\sec\left(\Theta - \frac{\pi}{2}\right) = 3\operatorname{cosec}\Theta$.

5. Write down the cartesian equations of the lines:

   (i) $r = 5 \sec\Theta$

   (ii) $r = 4\sec\left(\Theta - \frac{\pi}{2}\right)$

6. Write down the cartesian equations of the lines:

   (i) $r = 5\sec\left(\Theta - \frac{\pi}{3}\right)$

   (ii) $r = 7\sec\left(\Theta - \frac{\pi}{4}\right)$

Polar Curve Sketching

To sketch the polar curves, it is advisable to tabulate values of Θ in steps of 30° or $\frac{\pi}{6}$ in the range $0 \leq \Theta \leq 2\pi$ for a single angle Θ, such as $\sin \Theta$, $\cos \Theta$; steps of 15° or $\frac{\pi}{12}$ for double angles, such as $\sin 2\Theta$ and $\cos 2\Theta$; and steps of 10° or $\frac{\pi}{18}$ for treble angles, such as $\sin 3\Theta$ and $\cos 3\Theta$.

The student should know by heart the values of the sine and cosine of the angles in steps of 30° or $\frac{\pi}{6}$.

$\Theta°$	0	30	60	90	120	150	180
$\sin \Theta$	0	0.5	0.866	1	0.866	0.5	0
$\cos \Theta$	1	0.866	0.5	0	−0.5	−0.866	−1

contd …

$\Theta°$	210	240	270	300	330	360
$\sin \Theta$	−0.5	−0.866	−1	−0.866	−0.5	0
$\cos \Theta$	−0.866	−0.5	0	0.5	0.866	1

The work of tabulation can be reduced if the following points are taken into account.

(a) If r is a function of $\cos \Theta$, the curve is symmetrical about the initial line $\Theta = 0^c$ since $\cos(-\Theta) = \cos \Theta$ is an even function, so values of Θ up to and including π should only be recorded.

(b) If r is a function of $\sin \Theta$, the curve is symmetrical about the half-line $\Theta = \frac{\pi}{2}$ since $\sin(-\Theta) = -\sin \Theta$, is an odd function and the range used is $-\frac{\pi}{2} \leq \theta \leq \frac{\pi}{2}$.

In general, the student is advised to think in degrees and work in radians.

WORKED EXAMPLE 33

Plot the following polar coordinates on the same diagram.

$\mathbf{A}\left(3, \frac{\pi}{4}\right)$, $\mathbf{B}\left(4, \frac{\pi}{2}\right)$, $\mathbf{C}\left(-2, \frac{\pi}{4}\right)$, $\mathbf{D}(0, \pi)$.

Calculate the distances **AB, BC, BD** and **AC**.

Solution 33

The points **A, B, C** and **D** is plotted as shown in Fig. 5-I/64.

Using the cosine rule for the $\triangle$ **OAB**.

$AB^2 = OA^2 + OB^2 - 2(OA)(OB) \cos A\hat{O}B$
$= 3^2 + 4^2 - 2(3)(4) \cos \frac{\pi}{4} = 8.032$

$AB = 2.83$ units.

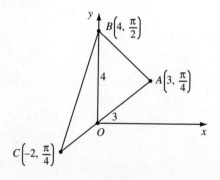

Fig. 5-I/64

From $\triangle BOC$ $BC^2 = 2^2 + 4^2 - 2(2)(4) \cos 135°$
$= 4 + 16 + 16(0.707) = 31.312$

$BC = 5.6$ units. $BD = OB = 4$ units $AC = 2 + 3 = 5$ units.

Worked Example 34

The diagonals of a rhombus **ABCD** intersect at the pole, if the polar coordinates of A and B are $(1, 0^c)$ and $\left(4, \frac{\pi}{2}\right)$ respectively, determine the following:-

(i) The polar coordinates of the other two points C and D.

(ii) The distance AB.

(iii) The area of the rhombus.

(iv) If the rhombus is rotated clockwise by $\frac{\pi}{4}$, about the pole, find the polar coordinates of A, B, C and D, in the range $-\pi \leq \Theta \leq \pi$.

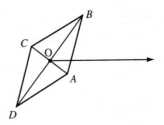

Fig. 5-I/66

Worked Example 35

Find the distance between the points $A\left(1, \frac{\pi}{8}\right)$ and $B\left(2, \frac{3\pi}{4}\right)$, and the area of the triangle OAB.

Solution 34

(i) The rhombus is drawn in Fig. 5-I/65. The diagonals are perpendicular to each other, $OC = OA = 1$, $C(-1, 0^c)$ or $C(1, \pi)$ $OB = OD = 4$,

$$D\left(-4, \frac{\pi}{2}\right) \text{ or } D\left(4, \frac{3\pi}{2}\right) \text{ or } D\left(4, -\frac{\pi}{2}\right).$$

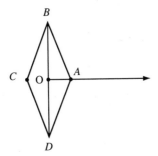

Fig. 5-I/65

(ii) $AB^2 = OA^2 + OB^2 = 1^2 + 4^2 = 17$
$AB = \sqrt{17}$ units.

(iii) Area of the rhombus $= 4 \times \frac{1}{2}(1 \times 4) = 8$ square units

(iv) The new values of the coordinates are:-

$A\left(1, -\frac{\pi}{4}\right)$, $B\left(4, \frac{\pi}{4}\right)$, $C\left(1, \frac{3\pi}{4}\right)$

and $D\left(4, -\frac{3\pi}{4}\right)$.

Fig. 5-I/66 shows the new position of the rhombus.

Solution 35

The points A and B are plotted as shown in the diagram of Fig. 5-I/67. The $\triangle OAB$ is constructed whose sides OA and OB are 1 and 2 respectively, the angle

$A\hat{O}B = \frac{3\pi}{4} - \frac{\pi}{8} = \frac{5\pi}{8}$ using the cosine rule

$AB^2 = OA^2 + OB^2 - 2(OA)(OB)\cos\frac{5\pi}{8}$

$= 1^2 + 2^2 - 2(1)(2)(-0.383)$

$= 6.532$

$AB = 2.56 \text{ units}.$

The distance between the points, $AB = 2.56$ units.

The area of the triangle $= \frac{1}{2}(OA)(OB)\sin\frac{5\pi}{8}$

$= \frac{1}{2}(1)(2)(0.924)$

$= 0.924$ square units.

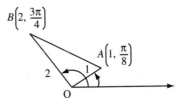

Fig. 5-I/67

Exercises 9

Draw the graphs of the following polar curves:-

(i) $r = 5 \sin \Theta$

(ii) $r = 4 \cos \Theta$

(iii) $r = 2 + 3 \cos \Theta$

(iv) $r = 3 + 2 \cos \Theta$

(v) $r = 2 - 3 \cos \Theta$

(vi) $r = 3 - 2 \cos \Theta$

(vii) $r = 4(1 + \cos \Theta)$

(viii) $r = 4(1 - \cos \Theta)$

(ix) $r = 5(1 + \sin \Theta)$

(x) $r = 5(1 - \sin \Theta)$

(xi) $r = 2 - 3 \sin \Theta$

(xii) $r = 2 + 3 \sin \Theta$

(xiii) $r = 3 + 2 \sin \Theta$

(xiv) $r = 3 - 2 \sin \Theta$

(xv) $r = 4 \sin 2\Theta$

(xvi) $r = 5 \cos 2\Theta$

(xvii) $r = 3 \cos 3\Theta$

(xviii) $r = 5 \sin 3\Theta$

(xix) $r^2 = 4 \sin 2\Theta$

(xx) $r^2 = 4 \cos 2\Theta$

(xxi) $r = 3\Theta$

(xxii) $r = -3\Theta$

(xxiii) $r = 3e^{\Theta}$

(xxiv) $r = 5$

(xxv) $r = \dfrac{1}{1 - \cos \Theta}$

(xxvi) $r = 5 \sin \Theta \tan \Theta$ cissoid of Diocles.

10

Area of the Triangle OPQ where $P(r_1, \Theta_1)$ and $Q(r_2, \Theta_2)$

The triangle OPQ is shown in Fig. 5-I/68
$P\hat{O}Q = \Theta_1 - \Theta_2$.

Area of $\triangle OPQ = \frac{1}{2}(OP)(OQ)\sin(\Theta_1 - \Theta_2)$

Area of $\triangle OPQ = \frac{1}{2}r_1 r_2 \sin(\Theta_1 - \Theta_2)$.

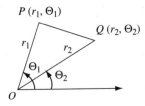

Fig. 5-I/68

WORKED EXAMPLE 36

Find the area of $\triangle OPQ$ where $P\left(1, \frac{\pi}{4}\right)$ and $Q\left(3, \frac{\pi}{3}\right)$.

Solution 36

Area of $\triangle OPQ = \frac{1}{2}(1)(3)\sin\left(\frac{\pi}{3} - \frac{\pi}{4}\right) = \frac{3}{2}\sin\frac{\pi}{12}$

$= 0.388$ square units.

WORKED EXAMPLE 37

Calculate the areas of the triangles OPQ, OPR, OQR and PQR. The coordinates of P, Q and R are given $\left(3, \frac{\pi}{2}\right)$, $\left(5, \frac{\pi}{3}\right)$ and $\left(4, \frac{\pi}{6}\right)$ respectively.

Solution 37

Area $\triangle OPQ = \frac{1}{2}(3)(5)\sin\frac{\pi}{6}$

$= 3.75$ square units

Area $\triangle OPR = \frac{1}{2}(3)(4)\sin\frac{\pi}{3}$

$= 3\sqrt{3}$

$= 5.196$ square units

Area $\triangle OQR = \frac{1}{2}(4)(5)\sin\frac{\pi}{6}$

$= 5$ square units

Area $\triangle PQR = $ Area $\triangle OPQ + $ Area $\triangle OQR$

$\qquad - $ Area $\triangle OPR$

$= 3.75 + 5 - 5.196$

$= 3.554$ square units.

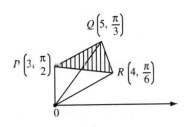

Fig. 5-I/69

Exercises 10

1. Mark the following points on a diagram:

   $P\left(5, \dfrac{\pi}{4}\right); Q\left(1, \dfrac{\pi}{2}\right); R\left(4, \dfrac{\pi}{6}\right).$

   Determine the following areas:

   (i) $\triangle OPQ$ (ii) $\triangle OPR$.

2. Mark the following points on a diagram

   $T\left(-5, \dfrac{\pi}{2}\right), S\left(6, \dfrac{\pi}{4}\right)$

   Determine the area of the triangle OTS

3. Find the exact area of the triangles
   (i) ORQ,   (ii) OQP,   (iii) ORP,

   given the points $R(5, \pi)$, $Q\left(4, \dfrac{\pi}{4}\right)$, $P\left(3, \dfrac{3\pi}{2}\right)$.

11

Polar to Cartesian and Vice-Versa

To Convert Polar to Cartesian Coordinates

WORKED EXAMPLE 38

Find the corresponding cartesian coordinates to the sets of polar coordinates and vice versa.

(i) $\left(3, \dfrac{\pi}{4}\right)$ (ii) $\left(2, \dfrac{\pi}{6}\right)$ (iii) $(5, 0)$

(iv) $\left(7, -\dfrac{\pi}{3}\right)$ (v) $\left(4, \dfrac{\pi}{12}\right)$.

Assuming a common origin and the x-axis as the initial line.

Solution 38

(a) (i) Fig. 5-I/70 shows the relationship between polar and cartesian.

$$\cos \Theta = \dfrac{x}{r} \text{ or } x = r \cos \Theta = 3 \cos \dfrac{\pi}{4} = 3\sqrt{2}$$

$$\sin \Theta = \dfrac{y}{r} \text{ or } y = r \sin \Theta = 3 \sin \dfrac{\pi}{4} = 3\sqrt{2}.$$

Fig. 5-I/70

Therefore $\left(3, \dfrac{\pi}{4}\right) \equiv \left(\dfrac{3}{\sqrt{2}}, \dfrac{3}{\sqrt{2}}\right)$.

(ii) $x = r \cos \Theta = 2 \cos \dfrac{\pi}{6} = 2\sqrt{\dfrac{3}{2}} = \sqrt{3}$

$y = r \sin \Theta = 2 \sin \dfrac{\pi}{6} = 2\left(\dfrac{1}{2}\right) = 1$.

Therefore $\left(2, \dfrac{\pi}{6}\right) \equiv \left(\sqrt{3}, 1\right)$

(iii) $x = r \cos \Theta = 5 \cos 0 = 5$

$y = r \sin \Theta = 5 \sin 0 = 0$.

Therefore, $(5, 0^c) \equiv (5, 0)$

(iv) $x = r \cos \Theta = 7 \cos \left(-\dfrac{\pi}{3}\right) = \dfrac{7}{2}$

$y = r \sin \Theta = 7 \sin \left(-\dfrac{\pi}{3}\right) = \dfrac{-7\sqrt{3}}{2}$.

Therefore, $\left(7, -\dfrac{\pi}{3}\right) \equiv \left(\dfrac{7}{2}, \dfrac{-7\sqrt{3}}{2}\right)$.

(v) $x = r \cos \Theta = 4 \cos \dfrac{\pi}{12} = 3.86$

$y = r \sin \Theta = 4 \sin \dfrac{\pi}{12} = 1.04$

Therefore, $\left(4, \dfrac{\pi}{12}\right) \equiv (3.86, 1.04)$.

To Convert Cartesian to Polar

(b) (i) Convert $\left(\dfrac{3}{\sqrt{2}}, \dfrac{3}{\sqrt{2}}\right)$ to polar coordinates

$x = r \cos \Theta$ ... (1)

$y = r \sin \Theta$ ... (2)

Dividing (2) by (1), we have $\tan \Theta = \dfrac{y}{x}$

$\dfrac{3\sqrt{2}}{3\sqrt{2}} = 1 \qquad \Theta = \tan^{-1} 1 = \dfrac{\pi}{4}$

39

$$r = \sqrt{x^2 + y^2} = \sqrt{\left(\frac{3}{\sqrt{2}}\right)^2 + \left(\frac{3}{\sqrt{2}}\right)^2} = 3.$$

Therefore, $\left(\frac{3}{\sqrt{2}}, \frac{3}{\sqrt{2}}\right) \equiv \left(3, \frac{\pi}{4}\right)$.

(ii) $(\sqrt{3}, 1)$ $\tan\Theta = \frac{y}{x} = \frac{1}{\sqrt{3}}$ or

$$\Theta = \tan^{-1}\frac{1}{\sqrt{3}} = \frac{\pi}{6},$$

$$r = \sqrt{(\sqrt{3})^2 + 1^2} = 2.$$

Therefore, $(\sqrt{3}, 1) \equiv \left(2, \frac{\pi}{6}\right)$.

(iii) $(5, 0)$ $\tan\Theta = \frac{y}{x} = \frac{0}{5} = 0$

or $\Theta = \tan^{-1} 0 = 0° = 0^c$ $r = \sqrt{5^2 + 0^2} = 5.$

(iv) $\left(\frac{7}{2}, \frac{-7\sqrt{3}}{2}\right)$

$$\tan\Theta = \frac{y}{x} = \frac{-\frac{7\sqrt{3}}{2}}{\frac{7}{2}} = -\sqrt{3}$$

or $\Theta = \tan^{-1} -\sqrt{3} = -\frac{\pi}{3}$

$$r = \sqrt{\left(\frac{7}{2}\right)^2 + \left(\frac{-7\sqrt{3}}{2}\right)^2} = 7.$$

Therefore, $\left(\frac{7}{2}, \frac{-7\sqrt{3}}{2}\right) \equiv \left(7, -\frac{\pi}{3}\right)$.

(v) $(3.86, 1.04)$ $\tan\Theta = \frac{y}{x} = \frac{1.04}{3.86}$

or $\Theta = 15°$ approximately

$r\cos\Theta = 3.86$ $r = \frac{3.86}{\cos 15°} = 4$

WORKED EXAMPLE 39

Sketch the following polar curves:-

(i) $r = \sin\Theta$ (ii) $r = \cos\Theta$
(iii) $r = 1 + \cos\Theta$ (iv) $r = 1 - \cos\Theta$
(v) $r = 1 + \sin\Theta$ (vi) $r = 1 - \sin\Theta$
(vii) $r = 1 + 2\cos\Theta$ (viii) $r = 2 + \cos\Theta$
(ix) $r = 1 - 2\sin\Theta$ (x) $r = 2 - \sin\Theta$.

Solution 39

(i) $r = \sin\Theta$

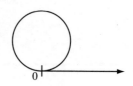

Fig. 5-I/71

(ii) $r = \cos\Theta$

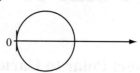

Fig. 5-I/72

(iii) $r = 1 + \cos\Theta$

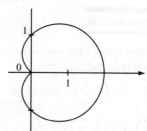

Fig. 5-I/73

(iv) $r = 1 - \cos\Theta$

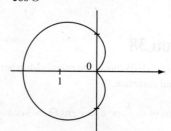

Fig. 5-I/74

(v) $r = 1 + \sin\Theta$

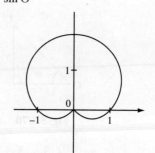

Fig. 5-I/75

(vi) $r = 1 - \sin \Theta$

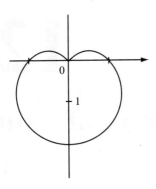

Fig. 5-I/76

(vii) $r = 1 + 2 \cos \Theta$

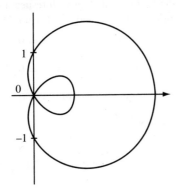

Fig. 5-I/77

(viii) $r = 2 + \cos \Theta$

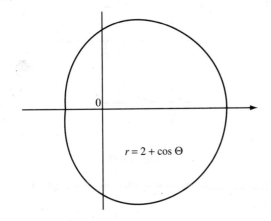

Fig. 5-I/78

(ix) $r = 1 - 2 \sin \Theta$

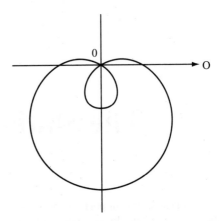

Fig. 5-I/79

(x) $r = 2 - \sin \Theta$

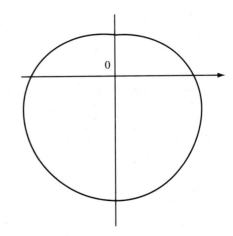

Fig. 5-I/80

Exercises 11

1. Determine the corresponding cartesian coordinates to the sets of polar coordinates and vice-versa.

   (i) $\left(4, \dfrac{\pi}{3}\right)$ (ii) $\left(5, \dfrac{\pi}{2}\right)$ (iii) $\left(10, \dfrac{-\pi}{4}\right)$

2. Convert $\left(\dfrac{1}{\sqrt{2}}, \dfrac{-1}{\sqrt{2}}\right)$ to polar coordinates and vice-versa.

12

The Stationary Values of $r \sin \Theta$

$r \sin \Theta$ is a general line in polar coordinates, which is parallel to the initial line. In cartesian coordinates $r \sin \Theta$ is equal to y.

Let $r = f(\Theta)$ be a curve in polar coordinates, the gradient at any point is $\frac{dr}{d\Theta} = f'(\Theta)$.

It is required to find the equation of a tangent or tangents to the curve $r = f(\Theta)$ which is or are parallel to the initial line.

The stationary values or the turning points or maxima and minima may be found if we differentiate $r \sin \Theta$ with respect to Θ.

$$\frac{d}{d\Theta}(r \sin \Theta) = \frac{dr}{d\Theta} \sin \Theta + r \cos \Theta$$ where r and Θ are variables.

For stationary values $\frac{dr}{d\Theta} \sin \Theta + r \cos \Theta = 0$

but $\frac{dr}{d\Theta} = f'(\Theta)$ and $r = f(\Theta)$ then

$$\boxed{f'(\Theta) \sin \Theta + f(\Theta) \cos \Theta = 0} \quad \ldots(1)$$

from which we can find the coordinates of the points at which the equations of the tangents are parallel to the initial line.

WORKED EXAMPLE 40

By considering the stationary values of $r \sin \Theta$ find the polar coordinates of the points on the curve $r = 1 + \cos \Theta$ where the tangent or tangents is or are parallel to the initial line and hence find the corresponding equations of the tangents.

Solution 40

$r = 1 + \cos \Theta$ ... the polar equation which is shown in Fig. 5-I/82. Differentiating with respect to Θ. $\frac{dr}{d\Theta} = -\sin \Theta$, substituting this and $r = 1 + \cos \Theta$ in (1)

$(-\sin \Theta) \sin \Theta + (1 + \cos \Theta) \cos \Theta = 0$

$-\sin^2 \Theta + \cos \Theta + \cos^2 \Theta = 0$

$2 \cos^2 \Theta + \cos \Theta - 1 = 0$

$$\cos \theta = \frac{-1 \pm \sqrt{1+8}}{4} = \frac{-1 \pm 3}{4}$$

$\cos \theta = -1, \theta = \pi; \quad \cos \theta = \frac{1}{2}, \quad \theta = \frac{\pi}{3}$ or $-\frac{\pi}{3}$.

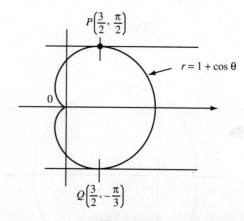

Fig. 5-I/81

It is seen clearly from Fig. 5-I/82 that the required points are $P\left(\frac{3}{2}, \frac{\pi}{3}\right)$ and $Q\left(\frac{3}{2}, -\frac{\pi}{3}\right)$ since

$r = 1 + \cos \Theta = 1 + \cos 60° = 1 + \frac{1}{2} = \frac{3}{2}$

$r = \frac{3}{2}$ when $\Theta = \frac{\pi}{3}$

$r = 1 + \cos \Theta = 1 + \cos(-60°) = \dfrac{3}{2}$

$r = \dfrac{3}{2}$ when $\Theta = -\dfrac{\pi}{3}$.

The equation of the tangent at $P\left(\dfrac{3}{2}, \dfrac{\pi}{3}\right)$

$r \sin \Theta = \left(1 + \cos \dfrac{\pi}{3}\right) \sin \dfrac{\pi}{3}$ or $r \sin \Theta = \dfrac{3}{2} \dfrac{\sqrt{3}}{2}$

$r = \dfrac{3\sqrt{3}}{4} \operatorname{cosec} \Theta = \dfrac{3\sqrt{3}}{2} \sec\left(\Theta - \dfrac{\pi}{2}\right)$

$$\boxed{r = \dfrac{3\sqrt{3}}{4} \sec\left(\Theta - \dfrac{\pi}{2}\right)}$$

The equation of the tangent at $Q\left(\dfrac{3}{2}, -\dfrac{\pi}{3}\right)$

$r \sin \Theta = \left(1 + \cos\left(-\dfrac{\pi}{3}\right)\right) \sin\left(-\dfrac{\pi}{3}\right) r \sin \Theta$

$= \dfrac{3}{2} \cdot \left(-\dfrac{\sqrt{3}}{2}\right) = -\dfrac{3\sqrt{3}}{4}$

$$\boxed{r = -\dfrac{3\sqrt{3}}{4} \sec\left(\Theta - \dfrac{\pi}{2}\right)}$$

WORKED EXAMPLE 41

Find the equation of the tangent, which is parallel to the initial line, on the curve $r = \cos \Theta$ in the range $0° \le \Theta \le \dfrac{\pi}{2}$. You may consider the stationary values of $r \sin \Theta$.

Solution 41

The derivative of $r \sin \Theta$ is

$\dfrac{d}{d\Theta}(r \sin \Theta) = \dfrac{dr}{d\Theta} \sin \Theta + r \cos \Theta$

For stationary values $\dfrac{dr}{d\Theta} \sin \Theta + r \cos \Theta = 0$.

The curve is $r = \cos \Theta$ then $\dfrac{dr}{d\Theta}$

$= -\sin \Theta(-\sin \Theta)(\sin \Theta)$

$+ \cos \Theta \cos \Theta = 0$

$\cos^2 \Theta - \sin^2 \Theta = 0 \quad \cos 2\Theta = 0 = \cos \dfrac{\pi}{2} \quad \Theta = \dfrac{\pi}{4}$.

The coordinates of P are $r = \cos \Theta = \cos \dfrac{\pi}{4} = \dfrac{1}{\sqrt{2}}$

and $\Theta = \dfrac{\pi}{4}$, $P\left(\dfrac{1}{\sqrt{2}}, \dfrac{\pi}{4}\right)$.

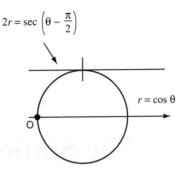

Fig. 5-I/82

Fig. 5-I/82 shows the polar curve $r = \cos \Theta$.

The equation of the tangent at $P\left(\dfrac{1}{\sqrt{2}}, \dfrac{\pi}{4}\right)$

$r \sin \Theta = \cos \Theta \sin \Theta = \cos \dfrac{\pi}{4} \sin \dfrac{\pi}{4} = \dfrac{1}{\sqrt{2}} \dfrac{1}{\sqrt{2}} = \dfrac{1}{2}$

$r = \dfrac{1}{2} \operatorname{cosec} \Theta \qquad \boxed{2r = \sec\left(\Theta - \dfrac{\pi}{2}\right)}$

Exercises 12

By considering the stationary values of $r \sin \Theta$, find the polar coordinates of the points on the following curves where the tangent or tangents is or are parallel to the initial line, and hence find the corresponding equations of the tangents:-

(i) $r = 5 \sin \Theta$

(ii) $r = 4 \cos \Theta$

(iii) $r = 5(1 + \cos \Theta)$

(iv) $r = 2 - 3 \cos \Theta$

(v) $r = 2 + 3 \cos \Theta$

(vi) $r = 3 - 2 \cos \Theta$

(vii) $r = 3 + 2 \cos \Theta$

(viii) $r = 5(1 - \sin \Theta)$

(ix) $r = 4(1 - \cos \Theta)$

(x) $r = 4(1 + \cos \Theta)$

(xi) $r = 2 - 3 \sin \Theta$

(xii) $r = 2 + 3 \sin \Theta$

(xiii) $r = 3 + 2 \sin \Theta$

(xiv) $r = 3 - 2 \sin \Theta$

(xv) $r = 5 \cos 2\Theta$

(xvi) $r = 3 \cos 3\Theta$

(xvii) $r = 5$

(xviii) $r^2 = 4 \cos 2\Theta$.

13

The Stationary Values of $r \cos \Theta$

$r \cos \Theta$ is a general line perpendicular to the initial line, in cartesian coordinates $r \cos \Theta$ is equal to x.

The stationary values are found by differentiating the function $r \cos \Theta$ with respect to $\Theta \frac{d}{d\Theta}(r \cos \Theta) = \frac{dr}{d\Theta} \cos \Theta - r \sin \Theta = 0 \ldots$ for stationary values.

Exercises 13

By considering the stationary values of $r \cos \Theta$, find the polar coordinates of the points on the following curves where the tangent or tangents is or are perpendicular to the initial line, and hence find the corresponding equations of the tangents

(i) $r = 5 \sin \Theta$

(ii) $r = 2 + 3 \cos \Theta$

(iii) $r = 4(1 + \cos \Theta)$

(iv) $r = 5 \cos 2\Theta$

where $0° \leq \Theta \leq 360°$.

14

Areas of Polar Curves

Area of a Sector

Consider the radius vector, $OP = r$ and the vectorial angle (positive), Θ. Let r be increased to $r + \delta r$ and Θ to $\Theta + \delta\Theta$ as shown in Fig. 5-I/84.

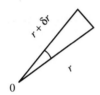

Fig. 5-I/83

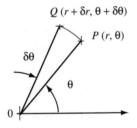

Fig. 5-I/84

The area of the sector OPQ is approximately equal to the area of the triangle POQ

$$\delta A \approx \frac{1}{2} r(r + \delta r) \sin \delta\Theta \qquad \frac{\delta A}{\delta\Theta} \approx \frac{1}{2} r(r + \delta r) \frac{\sin \delta\Theta}{\delta\Theta}$$

Fig. 5-I/85 shows the triangle POQ as $\delta\Theta \to 0$, $\delta r \to 0$,

$$\frac{\sin \delta\Theta}{\delta\Theta} \to 1 \text{ and } \frac{\delta A}{\delta\Theta} \to \frac{dA}{d\Theta}. \text{ Therefore, } \frac{dA}{d\Theta} = \frac{1}{2} r^2.$$

The area enclosed by the two half-lines $\Theta = \alpha$ and $\Theta = \beta$ is therefore given by the integral

$$\text{Area} = \frac{1}{2} \int_{\Theta=\alpha}^{\Theta=\beta} r^2 \, d\Theta \text{ and the length of arc.}$$

$$\text{Length of Arc} = \int_{\Theta=\alpha}^{\Theta=\beta} \sqrt{r^2 + \left(\frac{dr}{d\Theta}\right)^2} \, d\Theta.$$

WORKED EXAMPLE 42

Sketch the curve, where r and Θ are polar coordinates, $r = 1 + 2\cos\Theta$. Calculate the area contained in the inner loop.

Solution 42

Using the table, sketch the curve $r = 1 + 2\cos\Theta$ as shown in Fig. 5-I/85.

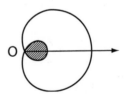

Fig. 5-I/85

$\Theta°$	0	30	60	90	120	150	180
$\cos\Theta$	1	0.866	0.5	0	−0.5	−0.866	−1
$2\cos\Theta$	2	1.732	1	0	−1	−1.732	−2
$r = 1 + 2\cos\Theta$	3	2.732	2	1	0	−0.732	−1

$\Theta°$	210	240	270	300	330	360
$\cos\Theta$	−0.866	−0.5	0	0.5	0.866	1
$2\cos\Theta$	−1.732	−1	0	1	1.732	2
$r = 1 + 2\cos\Theta$	−0.732	0	1	2	2.732	3

The limits of the integration are $\frac{2\pi}{3}$ and π.

The area contained in the inner loop

$$= 2 \times \frac{1}{2} \int_{\frac{2\pi}{3}}^{\pi} (1 + 2\cos\Theta)^2 \, d\Theta.$$

The area required $= \int_{\frac{2\pi}{3}}^{\pi} (1 + 2\cos\Theta)^2 \, d\Theta$

$$= \int_{\frac{2\pi}{3}}^{\pi} (1 + 4\cos^2\Theta + 4\cos\Theta) \, d\Theta$$

$$= \int_{\frac{2\pi}{3}}^{\pi} [1 + 2(\cos 2\Theta + 1) + 4\cos\Theta] \, d\Theta$$

where $\cos^2\Theta = \frac{\cos 2\Theta + 1}{2}$.

$$= \left[\Theta + \sin 2\Theta + 2\Theta + 4\sin\Theta \right]_{\frac{2\pi}{3}}^{\pi}$$

$$= 3\pi - \frac{2\pi}{3}(3) + \sin 2\pi - \sin \frac{4\pi}{3}$$

$$+ 4\sin\pi - 4\sin\frac{2\pi}{3}$$

$$= \pi + \frac{\sqrt{3}}{2} - 4\frac{\sqrt{3}}{2}$$

$$= \pi - \frac{3\sqrt{3}}{2}.$$

Exercises 14

1. Find the areas of the inner loops of the following polar curves:-
   (i) $r = 2 - 3\cos\Theta$
   (ii) $r = 2 + 3\cos\Theta$
   (iii) $r = 2 - 3\sin\Theta$
   (iv) $r = 2 + 3\sin\Theta$.

2. Find the area of one loop of the following polar curves:-
   (i) $r = 4\sin 2\Theta$
   (ii) $r = 5\cos 2\Theta$
   (iii) $r = 5\sin 3\Theta$
   (iv) $r = 3\cos 3\Theta$.

3. Find the small area enclosed by the curves $r = 3\Theta$ and $r = -3\Theta$.

4. Find the area enclosed by the curve $r = 5\sin\Theta$ and the two half-lines $\Theta = \frac{\pi}{3}$ and $\Theta = \frac{2\pi}{3}$.

5. Find the area enclosed by the curve $r = 4\cos\Theta$ and the initial line.

6. Find the area enclosed by the curve $r = 5(1+\sin\Theta)$ and the initial line.

7. Find the area enclosed by the curve $r = 3 - 2\cos\Theta$ and the line $2r = \sec\Theta$.

8. Find the area enclosed by the curve $r = 4(1 + \cos\Theta)$ and the line $r = 3\sec\Theta$.

9. Find the area enclosed by the half-lines $\Theta = \frac{\pi}{2}$ and $\Theta = \frac{\pi}{4}$ and the line $r = 2\sec\left(\Theta - \frac{\pi}{4}\right)$.

10. Find the total area of the loops of the following curves:-
    (i) $r^2 = a^2 \cos 2\Theta$
    (ii) $r^2 = a^2 \sin 2\Theta$.

Miscellaneous

1. Sketch the curve with equation $r = a\cos 2\Theta$, where $a > 0$ for $-\frac{\pi}{4} \leq \Theta \leq \frac{\pi}{4}$, and find the area of the region for which $0 \leq r \leq a\cos 2\Theta$, $-\frac{\pi}{4} \leq \Theta \leq \frac{\pi}{4}$. By considering the stationary values of $\sin\Theta \cos 2\Theta$, or otherwise, find the polar coordinates of the points on the curve where the tangent is parallel to the initial line.

   $\left(\text{Ans.}\,\frac{\pi a^2}{8},\,\left(\frac{2a}{3},\,\pm\sin^{-1}\frac{1}{\sqrt{6}}\right)\right)$

2. Sketch on the same diagram the curves whose polar equations are

   (i) $r = a(1+\cos\Theta)$, for $-\pi \leq \Theta \leq \pi$.

   (ii) $r = a\cos\Theta$ for $-\frac{\pi}{2} \leq \Theta \leq \frac{\pi}{2}$,

   where a is a positive constant. Calculate the area of the region which lies within the first curve but outside the second curve. The half-line $\Theta = \beta$, where $-\frac{\pi}{2} < \beta < \frac{\pi}{2}$, meets the curves at the pole O and at P and Q. Determine the polar equation of the locus of the mid-point of PQ as β varies.

   $\left(\text{Ans. Area} = \frac{5\pi a^2}{4};\,r = a\cos\Theta + \frac{a}{2}\right)$

3. Sketch on the same diagram the curves whose equations in polar coordinates are $r = 2a(1+\cos\Theta)$ and $r = a(3+2\cos\Theta)$, where a is a positive constant. Find the area of the region within which $2a(1+\cos\Theta) < r < a(3+2\cos\Theta)$.

   (Ans. $5\pi a^2$)

4. Find a polar equation of the curve $x^2 + y^2 = 4x$ and calculate the polar coordinates of the two points P and Q where the curve intersects the line

   $r = 2\sqrt{2}\sec\left(\frac{\pi}{4} - \Theta\right)$.

   Find the polar equations of the two half-lines from the origin which are tangents to the circle which has PQ as diameters.

   $\left(\text{Ans.}\,r = 4\cos\Theta;\,\left(2\sqrt{2},\,\frac{\pi}{4}\right);\right.$
   $\left.(4,0);\,\Theta = \frac{\pi}{4};\,\Theta = \frac{\pi}{4} - 2\tan^{-1}\frac{1}{2}\right)$.

5. Sketch the curve with polar equation $r = 1 + \cos\Theta$ and find the value of Θ, where $0 < \Theta < \pi$, at the point P where $r = \frac{3}{2}$. Verify that the line with equation $r = \frac{3\sqrt{3}}{4}\sec\left(\frac{\pi}{2} - \Theta\right)$ is a tangent to the curve at P, and draw this tangent on your sketch. Find the polar equation of the tangent which is parallel to the half-line $\Theta = \frac{\pi}{2}$ and touches the curve at $S(r,\alpha)$ where $-\frac{\pi}{2} < \alpha < \frac{\pi}{2}$.

   If the tangents at S and P meet at T and O is the pole, find the polar equation of the half-line TO and show that the values of r at the two points in which TO or TO produced meet the curve again are $1 \pm \frac{8}{\sqrt{91}}$.

6. Sketch, for $0 \leq \Theta \leq 4\pi$, the curve C_1 whose equation in polar coordinates is $r = 3\Theta$, marking on your sketch the polar coordinates of the points of intersection with the initial line and the half-lines $\Theta = \frac{\pi}{2}$, $\Theta = \pi$, $\Theta = \frac{3\pi}{2}$.

   Calculate the area of the region enclosed by the portion of the curve C_1 for which $0 \leq \Theta \leq \frac{\pi}{2}$, and the half-line $\Theta = \frac{\pi}{2}$. On the same diagram sketch the curve C_2 whose equation is $r = \frac{(3\pi\sin\Theta)}{2}$, $0 \leq \Theta \leq \pi$, and calculate the area of the region enclosed between the curves C_1 and C_2 for the range $0 \leq \Theta \leq \frac{\pi}{2}$. [If required, you may assume that $\sin\Theta > \frac{2\Theta}{\pi}$ for $0 < \Theta < \frac{\pi}{2}$].

   $\left(\text{Ans.}\,\frac{3\pi^3}{16},\,\frac{3\pi^3}{32}\right)$.

7. Draw a diagram to show the regions **A** and **B**, which are defined in terms of polar coordinates by $A = \{(r,\Theta): r < 4\cos\Theta\}$ and $B = \{(r,\Theta): r < 3 - 2\cos\Theta\}$. Show that the area of the region $A \cap B$ is $\frac{5(2\pi - 3\sqrt{3})}{2}$.

47

8. Sketch on the same diagrams, for $0 \le \Theta < 2\pi$, the curves C_1 and C_2 with polar equations $r = 2$ and $r = 2\sin 2\Theta$ respectively, where $r \ge 0$, showing in particular the behaviour of the curve C_2 near the pole. Find the area of the region which lies inside C_1 and outside C_2, leaving your answer in terms of π.

(Ans. 3π)

9. Sketch the graphs of the polar curves $r = 2\sec(\alpha - \Theta)$, where $0 < \alpha < \frac{\pi}{2}$ and $\tan \alpha = 2$, and $r = \left(\sqrt{5}\right)(1 + \cos \Theta)$. Verify that the points with polar coordinates $\left(2\sqrt{5}, 0\right)$ and $\left(\sqrt{5}, \frac{\pi}{2}\right)$ lie on both graphs.

Calculate the area of the region defined by the inequalities

$$2\sec(\alpha - \Theta) \le r \le \left(\sqrt{5}\right)(1 + \cos \Theta),$$

$$0 \le \Theta \le \frac{\pi}{2}.$$

$\left(\text{Ans.} \dfrac{15\pi}{8}\right)$

10. Sketch, on the same diagram the curves defined by the polar equations $r = a$ and $r = a(1 + \cos \Theta)$, where a is a positive constant and $-\pi < \Theta \le \pi$. By considering the stationary value of $r \sin \Theta$, or otherwise, find equations of the tangents to the curve $r = a(1 + \cos \Theta)$ which are parallel to the initial line. Show that the area of the region for which $a < r < a(1 + \cos \Theta)$ is $\dfrac{(\pi+8)a^2}{4}$.

11. The equations, in polar coordinates, of 3 curves C_1 C_2 and C_3 are respectively $r = a(1+\cos \Theta)$, $-\pi < \Theta \le \pi$; $r = 2a \cos \Theta$, $-\frac{\pi}{2} \le \Theta \le \frac{\pi}{2}$, and $r = \frac{3a}{4}\sec \Theta$, $-\frac{\pi}{2} < \Theta < \frac{\pi}{2}$, where $a > 0$.

   (a) Find the point, or points, which C_1 and C_2 have in common.

   (b) Find the point, or points, that C_1 and C_3 have in common.

   (c) Sketch the three curves on the same diagram.

   (d) Show that the area of the region inside C_1 and outside C_2 is $\frac{1}{2}\pi a^2$.

$\left(\text{Ans. (a) } r = 2a, \Theta = 0; r = 0\right.$

$\left.\text{(b) } r = \dfrac{3a}{2}, \Theta = \pm \dfrac{\pi}{3}\right).$

12. For $0 \le \Theta \le \frac{\pi}{2}$, sketch, on the same diagram, the curve C with polar equation $r = 4p\Theta$ and the line L with polar equation $r = \pi p \sec\left(\frac{\pi}{4} - \Theta\right)$, where p is a positive constant. At the point where $\Theta = \Theta_1$, the tangent to the curve C is perpendicular to the initial line. By considering the stationary values of $r \cos \Theta$, or otherwise, show that $\Theta_1 = \cot \Theta_1$. Show that C and L meet A the point $\left(p\pi, \frac{\pi}{4}\right)$ and find the area of the finite region bounded by C, L and the initial line.

$\left(\text{Ans.} \dfrac{\pi^2 p^2}{24}(12 - \pi)\right).$

13. Sketch the curve whose equation in polar coordinates is $r^2 = a^2 \cos 2\Theta$, $-\pi < \Theta \le \pi$. Where a is a positive constant. Using the fact that a tangent is parallel to the initial line when $r \sin \Theta$ is stationary or otherwise, find the polar coordinates of the points at which the tangents to the above curve are parallel to the initial line. Calculate the area enclosed by one loop of this curve.

Find also the area of the region which lies inside this loop but outside the curve

$$r = \dfrac{a}{\sqrt{2}}, -\pi < \Theta \le \pi.$$

$\left(\text{Ans.} \left(\dfrac{a}{\sqrt{2}}, \dfrac{\pi}{6}\right), \left(\dfrac{a}{\sqrt{2}}, \dfrac{-\pi}{6}\right), \left(\dfrac{a}{\sqrt{2}}, \dfrac{5\pi}{6}\right),\right.$

$\left.\left(\dfrac{a}{\sqrt{2}}, \dfrac{-5\pi}{6}\right) \cdot \dfrac{a^2}{2}, \dfrac{a^2}{2}\left(\sqrt{3} - \dfrac{\pi}{3}\right)\right).$

14. Sketch, for $0 \le \Theta \le 2\pi$, and on separate diagrams, the curves $C_1 C_2$ and C_3 whose polar equations are

   (a) $r = 4a$,

   (b) $r = 4a \cos 2\Theta$

   (c) $r = -4a \cos 2\Theta$

respectively, where a is positive constant and $r \ge 0$.

Mark on your sketch of C_3 the magnitudes of the angles between the tangents to the curve at the pole and the half-line $\Theta = 0°$.

Find the area of the region which lies inside C_1 but outside C_2 and C_3 when the three sketches are superimposed.

(Ans. Angles are 45°, 135°; $8\pi a^2$)

15. Sketch one the same diagram, the curves C_1, C_2 with respective polar equations $r = a\cos 2\Theta$, $r = a\cos\Theta$, where a is a positive constant and $r \geq 0$. By considering the stationary values of $\sin\Theta\cos 2\Theta$, or otherwise, find polar coordinates of the four points on C_1 where the tangent is parallel to the initial line. Find the area of the region which lies inside C_2 but other side C_1.

$\left(\text{Ans. } \left(\dfrac{2a}{3}, \pm 24.1^0\right),\right.$

$\left.\left(\dfrac{2a}{3}, \pm 155.9^0\right) \text{ Area} = \dfrac{\pi a^2}{8}\right)$

16. Sketch on the same diagram the curves C_1 and C_2 defined by the polar equations

    $C_1: r\cos\Theta = 2, \ -\dfrac{\pi}{2} < \Theta < \dfrac{\pi}{2},$

    $C_2: r = \dfrac{6}{1+\cos\Theta}, \ -\pi < \Theta < \pi.$

    (a) State the polar coordinates of the point of intersection of C_1 with the initial line.

    (b) State the polar coordinates of the points of intersction of C_2 with the initial line and the half lines $\Theta = \dfrac{\pi}{2}$ and $\Theta = -\dfrac{\pi}{2}$.

    (c) Find the polar coordinates of the points of intersection of C_1 and C_2.

    (d) Calculate the area of the finite region bounded by C_1 and C_2.

    $\left(\text{Ans. (a) } (2, 0),\right.$

    $(b) \ (3, 0), \left(6, \dfrac{\pi}{2}\right), \left(6, -\dfrac{\pi}{2}\right).$

    $\left.(c) \ \left(4, \dfrac{\pi}{3}\right), \left(4, -\dfrac{\pi}{3}\right), (d) \ \dfrac{8\sqrt{3}}{3}\right).$

17. Sketch on the same diagram the curves whose polar equations are $r = 3a\sec\Theta, \ -\dfrac{\pi}{2} < \Theta < \dfrac{\pi}{2}$, and $r = 4a(1+\cos\Theta), \ -\dfrac{\pi}{3} \leq \Theta \leq \dfrac{\pi}{3}$.

18. Sketch on the same diagram, the curves C_1, C_2 whose polar equations are $C_1: r = a(1+\cos\Theta)$, $-\pi < \Theta \leq \pi$. $C_2: r = b(1-\cos\Theta), \ -\pi < \Theta \leq \pi$ where a and b are positive constants with $b > a$.

    Find the points common to C_1 and C_2. Calculate the area of the finite region bounded by the arcs with polar equations $r = a(1+\cos\Theta)$,

    $\dfrac{\pi}{2} \leq \Theta \leq \pi, r = a(1-\cos\Theta), 0 \leq \Theta \leq \dfrac{\pi}{2}$.

19. The curves C_1 and C_2 have polar equations C_1: $r^2 = a^2\sin 2\Theta, \ 0 \leq \Theta < 2\pi$, $C_2: r = a\cos\Theta$, $-\dfrac{\pi}{2} < \Theta \leq \dfrac{\pi}{2}, a > 0$.

    (a) Sketch, on the same diagram, the curves C_1 and C_2.

    (b) Find the polar coordinates of the points of intersection of C_1 and C_2.

    (c) Show that the area of the finite region which is inside C_1 and is also inside C_2 is $\dfrac{a^2}{4}\arctan 2$.

20. The parametric equations of a curve C are $x = a(\Theta - \sin\Theta) \ \ y = a(1+\cos\Theta), 0 \leq \Theta \leq 6\pi$.

    (a) Show that the gradient of C at the point with parameter Θ is $-\cot\left(\dfrac{1}{2}\Theta\right)$.

    Sketch C showing clearly the coordinates of the points where C meets the axes.

    (b) Show that the length of C between $\Theta = 0$ and $\Theta = \pi$ is $4a$.

    (c) Find the area of the curved surface generated when the arc of C between $\Theta = 0$ and $\Theta = \pi$ is rotated through 2π about the y-axis.

21. Sketch, on the same diagram, the curves C_1, C_2 whose polar equations are $C_1: r = 5\cos\Theta, -\dfrac{1}{2}\pi \leq \Theta < \dfrac{1}{2}\pi$, $C_2: r = 2+\cos\Theta, -\pi \leq \Theta < \pi$.

    Find a polar equation of the straight line which passes through the points of intersection of C_1 and C_2.

    Calculate the area of the region which lies inside the curve C_1 and outside the curve C_2.

22. (i) Show that, for real x, $\dfrac{2}{3} \leq \dfrac{x^2+1}{x^2+x+1} \leq 2$.

    Find the equations of any asymptotes to the curve $y = \dfrac{x^2+1}{x^2+x+1}$.

    Hence, or otherwise, sketch the graph of

    $y = \dfrac{x^2+1}{x^2+x+1}$,

    stating clearly the coordinates of any maxima or minima.

    (ii) Find the set of values of for which $|2x-4| - |x+2| > 2$.

23. The curve with equation $y = \frac{ax+b}{x(x+2)}$, where a and b are constants, has zero gradient at the point $(1, -2)$.

   (a) Show that $a = -8$ and find the value of b. (6 marks)

   (b) Show that the gradient is also zero at the point $\left(-\frac{1}{2}, -8\right)$. (2 marks)

   (c) Find equations for the three asymptotes of this curve. (2 marks)

   Sketch the curve, stating the coordinates of the point at which the curve meets the x-axis. (3 marks)

   Using your sketch, or otherwise, find the set of values of y for which no part of the curve exists. (2 marks)

24. The curve C has equations $y = \frac{(x-3)^2}{x+1}$.

   (a) By considering the set of values of y for real x, show that no part of C exists in the interval $-16 < y < 0$. (4 marks)

   (b) Show that the line $y = x - 7$ is an asymptote to C and state the equation of the other asymptote.

   Sketch C showing the coordinates of points at which C meets the coordinate axes and the way in which C approaches the asymptotes. (3 marks)

   Show that the equation $(x+1)\ln x - (x-3)^2 = 0$ has a root x in the interval $1.6 < x < 2$. (1 mark)

   Taking 1.6 as a first approximation to a, use the Newton-Raphson method once to find a second approximtion giving your answer to two decimal places. (4 marks)

25. A curve has equation $y = \frac{x^2 - 1}{3x - 5}$.

   (a) Prove that, for real values of x, the value of y cannot lie between $\frac{2}{9}$ and 2. (5 marks)

   (b) Find the coordinates of the turning points of the curve. (3 marks)

   (c) Show that one asymptote to the curve has equation $9y = 3x + 5$ and state the equation of the other asymptote. (3 marks)

   (d) Sketch the curve, showing its asymptotes. (4 marks)

26. A curve has equation $y = x^3 - 6x^2 + 3x + 10$

   (a) Sketch this curve showing the coordinates of points at which the curve meets the coordinate axes. (4 marks)

   (b) Find the coordinates of the point of inflexion of the curve. (3 marks)

27. Given that x is real, show that $-4 \leq \frac{4x-3}{x^2+1} \leq 1$. (5 marks)

   Sketch the curve with equation $y = \frac{4x-3}{x^2+1}$ showing clearly on your sketch

   (a) the coordinates of the points where the curve crosses the coordinates axes (1 mark)

   (b) the coordinates of the maximum point and the minimum point (2 marks)

   (c) the shape of the curve for large values of $[x]$ (2 marks)

   Verify the equation $\frac{4x-3}{x^2+1} - e^{-x} = 0$ has a root betwen 0.9 and 1. (1 mark)

   Taking 1 as a first approximation to this root, apply the Newton-Raphson method once to find a second approximation giving your answer to two decimal places.

28. Given that x is real and $y = \frac{(x-2)^2}{x^2+4}$, show that $0 \leq y \leq 2$. (3 marks)

   Hence write down the coordinates of the two stationary points on the curve with equation $y = \frac{(x-2)^2}{x^2+4}$ (2 marks)

   Sketch the curve sketching how the curve approaches its asymptote. (3 marks)

   With the aid of your sketch, explain why the equation $x(x^2 + 4) = (x - 2)^2$ has only one real root. Verify, by calculation that this root lies betwen 0.5 and 0.6. (3 marks)

   Taking 0.5 as a first approximation to the real root of this equation, use the Newton-Raphson method once to determine a second approximation, giving your answer to 2 decimal places. (4 marks)

29. For the curve whose equation is $y = \frac{4}{x-4} - \frac{1}{x-1}$, find the coordinates of the turning points and the equations of the 3 asymptotes. (6 marks)

Sketch the curve and state the set of values of y for which no part of the curve exists. (4 marks)

Find the 2 non-zero values of m for which the line $y = mx$ is a tangent to the curve. Verify that these two tangents are perpendicular. (5 marks)

30. By converting from the polar coordinates (r, Θ) to cartesian coordinates, or otherwise, show that the equation $\frac{2a}{r} = \cos\Theta + 2\sin\Theta$, where a is a positive constant, represents a straight line ?

Sketch L and the closed curve C given by $r = a(1 + \cos\Theta)$ on the same diagram. Write down the polar coordinates of the points of intersection of L and C. (6 marks)

Find, in terms of π and a, the area of the region enclosed by C. (4 marks)

The line L partitions the region enclosed by C into two regions of area R_1 and R_2, where $R_1 < R_2$. Find the ratio $R_1:R_2$. (5 marks)

31. Given that $f(x) = 2x^3 - 5x^2 - 4x + 3$, find the stationary values of $f(x)$. Show that $f(x) = (x+1)(2x-1)(x-3)$.

Hence sketch the curve with equation $y = f(x)$, marking on your sketch the coordinates of the points where the curve crosses the coordinate axes. (5 marks)

(a) Solve the inequalities
   (i) $2x^3 - 5x^2 - 4x + 3 > 0$, (2 marks)
   (ii) $2e^{3x} - 5e^{2x} - 4e^x + 3 > 0$. (3 marks)

(b) Express $\cos 2x$ and $\cos 3x$ in terms of $\cos x$ only, and hence find the general solution in radians, in terms of π, of the equation $1 + \cos 3x = 5(\cos 2x + \cos x)$. (6 marks)

32. Fig. 5-M/1 shows a sketch of the part of the graph of $y = f(x)$ for $0 \le x \le 2a$.

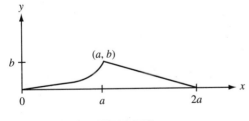

Fig. 5-M/1

The line $x = 2a$ is a line of symmetry of the graph.

Sketch on separate axes the graphs of

(a) $y = f(x)$ for $0 \le x \le 4a$. (2 marks)

(b) $y = -f(2x)$ for $0 \le x \le a$. (2 marks)

(c) $y = 3f\left(\frac{1}{2}x\right)$ for $0 \le x \le 2a$. (2 marks)

(d) $y = f(x - a)$ for $2a \le x \le 4a$. (2 marks)

33. A curve has cartesian equation $y^2 = \frac{x^2(4-x)}{(4+x)}$. State the equation of its asymptote. (2 marks)

Show that the equation of the curve can be expressed in the form $r = 4(2\cos\Theta)$ where (r, Θ) are polar coordinates. (4 marks)

Sketch the curve and prove that the area of the loop is $8(4 - \pi)$ (9 marks)

34.

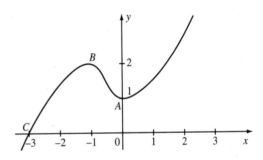

Fig. 5-M/2

The diagram shows the graph of $y = f(x)$. The points A, B, C have coordinates $(0, 1)$, $(-1, 2)$, $(-3, 0)$ respectively. Sketch separately the graphs of

(i) $y = f(-x)$

(ii) $y = f(x + 3)$,

showing in each case the coordinates of the points corresponding to A, B and C.

35. (a) A graph has polar equation
$r = \frac{2}{\cos\Theta \sin\alpha - \sin\Theta \cos\alpha}$, where α is a constant. Express the equation in cartesian form.

Hence sketch the graph in the case $\alpha = \frac{1}{8}\pi$, giving the cartesian coordinates of the intersections with the axes. [7]

(b) A graph has cartesian $(x^2 + y^2)^2 = 4x^2$. Express the equation in polar form. Hence, or otherwise, sketch the graph.

36. (a) Sketch the curve C_1 whose equation in polar coordinates is $r = a(1 + \sin\Theta)$, where a is a positive constant and $0 \leq \Theta < 2\pi$. [4]

Find the area of the region enclosed by C_1. [5]

(b) The curve C_2 has polar equation $r(1+\sin\Theta) = a$, where a is a positive constant. Show that the equation of C_2 may be expressed in cartesian form as $2ay = a^2 - x^2$, and sketch C_2.

37. The equation, in polar coordinates, of a curve C is $r = ae^{\frac{1}{2}\Theta}$, $0 \leq \Theta < 2\pi$, where a is a positive constant. Write down, in terms of Θ, the cartesian coordinates, x and y, of a general point P on the curve. Show that the gradient of the tangent at P is given by $\frac{dy}{dx} = \frac{\tan\Theta + 2}{1 - 2\tan\Theta}$. [6]

Hence show that the tangent at P is inclined to the radius vector OP at a constant angle α, where $\tan\alpha = 2$. [2]

Sketch the curve C. [3]

[An accurate graph is not required, but you should show clearly the range of values of r.]

Find the area of the region traced out by the radius vector OP as P moves on C from the point where $\Theta = 0$ to the point where $\Theta = \pi$.

5. CARTESIAN AND POLAR CURVE SKETCHING

Answers

Exercises 1

1. (0.168, 11.84) minimum.

2. (0, 0) minimum; $\left(\frac{\pi}{2}, 0\right)$ minimum;

   $\left(\frac{\pi}{4}, \frac{1}{2}\right)$ maximum; $(\pi, 0)$ minimum;

   $\left(\frac{3\pi}{4}, \frac{1}{2}\right)$ maximum; $\left(\frac{5\pi}{4}, \frac{1}{2}\right)$ maximum;

   $\left(\frac{7\pi}{4}, \frac{1}{2}\right)$ maximum; $(2\pi, 0)$ minimum.

3. (i) $(-2, 4)$ minimum   (ii) $(2, 3)$ minimum

   (iii) $(-3, 1)$ maximum   (iv) $(3, 1)$ maximum.

4. (i) $(1.36, -20.74)$ minimum;
   $(-2.7, 12.6)$ maximum

   (ii) $(0.577, -0.385)$ minimum,
   $(-0.577, 0.385)$ maximum.

5. $(3, -48)$ minimum, $\left(-\frac{5}{3}, 2.81\right)$ maximum.

6. –

7. (i) $\left(\frac{3}{4}, -6\frac{1}{8}\right)$ (ii) $\left(-\frac{1}{3}, \frac{4}{3}\right)$ maximum

   (iii) $(0.215, 0.887)$ minimum
   $(-1.549, 11.06)$ maximum

   (iv) $(1.639, -11.86)$ minimum,
   $(-0.305, -4.49)$ maximum.

8. (i) minimum
   (ii) maximum
   (iii) minimum, maximum
   (iv) minimum, maximum.

9. (i) $\left(-\frac{2}{3}, 2.26\right)$

   (ii) $\left(\frac{2}{3}, -\frac{221}{27}\right)$

10. (i) $\left(-\frac{1}{2}, \frac{3}{2}\right)$ min   (ii) $\left(\frac{1}{2}, \frac{7}{2}\right)$ max

    (iii) $\left(-\frac{3}{2}, -2\right)$ max   (iv) $\left(\frac{5}{2}, \frac{5}{4}\right)$ max.

11. 72, 72.   12. $y = 10$ m, $h = 5$ m.

13. $I = 6.67 A$, $P_{max} 40$ W.

14. $x = 5.313$ m, $h = 5.313$ m.

15. $x = 10.3$ min $x = 3.03$ max $V_{max} = 513.1$.

Exercises 2

1. –   2. –   3. –

4. (i) $e^{-3x} - 3xe^{-3x}$, $-6e^{-3x} + 9xe^{-3x}$

   (ii) $e^{2x} + 2xe^{2x}$, $4e^{2x} + 4xe^{2x}$.

5. (i) $6(2x - 3)^2 e^{-x^2} - 2xe^{-x^2}(2x - 3)^3$

   (ii) $4(x + 1)^3 e^{x^3} + 3x^2(x + 1)^4 e^{x^3}$.

6. –   7. –   8. –

9. (i) $3e^{-x} - 3xe^{-x}$

   (ii) $-6e^{-x} + 3xe^{-x}$ $e^{-x^2} - 2x^2 e^{-x^2}$,
   $-6xe^{-x^2} + 4x^3 e^{-x^2}$

   (iii) $-x^{-2} e^{x^2} + 2e^{x^2}$, $\frac{2e^{x^2}}{x^3} - \frac{2e^{x^2}}{x} + 4e^{x^2}$.

10. $x = \frac{1}{\sqrt{6}}$, max, $x = -\frac{1}{\sqrt{6}}$, min; $x = 0.707$,
    $x = -0.707$ and $x = 0$ are three points of inflexions.

Exercises 3
1. -0.049 2. -1.63
3. —   4. —
5. $\dfrac{2}{3}$ 6. $-\dfrac{1}{2}, 1$.

7. No turning points, the gradients are everywhere positive

8. $(0.68, 0.271)$ min, $(-1.48, -0.59)$ max.

Exercises 4
Graphs.

Exercises 5
1. —

2. (i) 2   (ii) 4.96   (iii) $2\sqrt{3}$ (iv) 0   (v) 8   (vi) 3

3. (i) $PQ^2 = 1.91$ (ii) $PQ^2 = 7$ (iii) 5.93
   (iv) $PQ^2 = 2$ (v) $PQ^2 = 8$

4. $\hat{A} = 63°\,26'$, $\hat{B} = 82°\,53'$

Exercises 6
1. (i) $r = \dfrac{ab}{b\cos\Theta + a\sin\Theta}$
   $= \dfrac{ab}{\sqrt{a^2+b^2}} \sec(\Theta - \alpha)$ where $\alpha = \tan^{-1}\dfrac{a}{b}$

   (ii) $r = \dfrac{-c}{\sqrt{m^2+1}} \sec(\Theta + \alpha)$ where $\alpha = \tan^{-1}\dfrac{1}{m}$

   (iii) $r = \dfrac{-c}{a\cos\Theta + b\sin\Theta}$

   (iv) $r = \dfrac{2}{2\cos\Theta + \sin\Theta}$

   (v) $r = \dfrac{6}{2\sin\Theta - 3\cos\Theta}$

   (vi) $r = \dfrac{-5}{\sin\Theta + 3\cos\Theta}$

   (vii) $r = \dfrac{3}{\sqrt{5}} \sec(\Theta - \alpha)$ where $\alpha = \tan^{-1} 2$

   (viii) $\Theta = \tan^{-1}(-3)$ (ix) $\Theta = \dfrac{\pi}{4}$

   (x) $r = 5\sec\Theta$ (xi) $r = -3\sec\Theta$

   (xii) $r = 2\sec\left(\dfrac{\pi}{2} - \Theta\right)$ (xiii) $5\sec\left(\Theta - \dfrac{\pi}{2}\right)$

   (xiv) $\Theta = 0$ (xv) $\Theta = \dfrac{\pi}{2}$

   (xiv) $r = \dfrac{1}{\sin\Theta - \cos\Theta}$

   (xvii) $r = \dfrac{-2}{\sin\Theta + \cos\Theta}$

   (xviii) $\Theta = \tan^{-1} 2$ (xix) $r = -10\sec\Theta$

   (xx) $r = \dfrac{1}{\sin\Theta + \cos\Theta}$

2. (i) $r = 1$ (ii) $r = 5$

   (iii) $r^2 = \dfrac{144}{16\cos^2\Theta + 9\sin^2\Theta}$

   (iv) $r^2 = \dfrac{400}{25\cos^2\Theta - 16\sin^2\Theta}$

   (v) $r^2 = 10\,\text{cosec}\,2\Theta$

   (vi) $r = 4\cos\Theta\,\text{cosec}^2\Theta$

   (vii) $r = -4\sin\Theta\,\sec^2\Theta$

   (viii) $r = \dfrac{20}{4\cos\Theta - r\sin^2\Theta}$

   (ix) $r = \dfrac{\sin\Theta \pm \sqrt{\sin^2\Theta - 4\cos^2\Theta}}{2\cos^2\Theta}$

   (x) $r = \dfrac{(3\cos\Theta + \sin\Theta \pm \sqrt{(3\cos\Theta + \sin\Theta)^2 + 4\cos^2\Theta})}{2\cos^2\Theta}$

   (xi) $r^2 = 2c^2\,\text{cosec}\,2\Theta$

   (xii) $r^2 = \dfrac{a^2 b^2}{b^2\cos^2\Theta + a^2\sin^2\Theta}$

   (xiii) $r^2 = a^2\sec 2\Theta$

   (xiv) $r = \dfrac{(-(10\sin\Theta - 6\cos\Theta) \pm \sqrt{(10\sin\Theta - 6\cos\Theta)^2 - 36})}{2}$

   (xv) $r = \dfrac{(\cos\Theta + \sin\Theta) \pm \sqrt{(\cos\Theta + \sin\Theta)^2 + 4}}{2}$

(xvi) $r^2 = 2 \csc 2\Theta$

(xvii) $r^2 = -2c^2 \csc 2\Theta$

3. (i) $r^2 = 2(x^2 - y^2)$

(ii) $(x^2 + y^2)^{\frac{3}{2}} = 6xy$

(iii) $(x^2 + y^2)^2 + 15x(x^2 + y^2) = 20x^3$

(iv) $(x^2 + y^2)^2 = 12(x^2 + y^2) - 16y^3$

(v) $x^2 + y^2 = \sqrt{2(x^2 + y^2)} + x$

(vi) $x^2 + y^2 = \sqrt{3}(x^2 + y^2)^{\frac{1}{2}} - y$

(vii) $x^2 + y^2 = (x^2 + y^2)^{\frac{1}{2}} + x$

(viii) $x^2 + y^2 = (x^2 + y^2)^{\frac{1}{2}} + y$

(ix) $x^2 + y^2 = (x^2 + y^2)^{\frac{1}{2}} - x$

(x) $x^2 + y^2 = (x^2 + y^2)^{\frac{1}{2}} - y$

(xi) $(x^2 + y^2)^2 = 4xy$

(xii) $(x^2 + y^2)^2 = 3(x^2 - y^2)$

(xiii) $x^2 + y^2 = 3(x^2 + y^2)^{\frac{1}{2}} + x$

(xiv) $x^2 + y^2 = (x^2 + y^2)^{\frac{1}{2}} + 2x$

(xv) $\dfrac{y}{x} = \tan \dfrac{\sqrt{x^2 + y^2}}{5}$

(xvi) $\dfrac{y}{x} = \tan \dfrac{\sqrt{x^2 + y^2}}{-3}$

(xvii) $y = \sqrt{3}x$

(xviii) $y = -3.08x$

(xix) $y = 3$

(xx) $x = 5$

(xxi) $y = 1$

(xxii) $y = 3$

(xxiii) $x^2 + y^2 = 3$

(xxiv) $x^2 + y^2 - 3y = 0$

(xxv) $x^2 + y^2 - 5x = 0$

(xxvi) $(x^2 + y^2)^2 = a^2(x^2 - y^2)$

(xxvii) $(x^2 + y^2)^2 = 2xya^2$

(xxviii) $(x^2 + y^2)^{\frac{5}{2}} = (x^2 - y^2)^2$

(xxix) $(x^2 + y^2)^{\frac{5}{2}} = 4x^2 y^2$

(xxx) $(x^2 + y^2)^{\frac{1}{2}} = \dfrac{3}{x \cos\alpha + y \sin\alpha}$

(xxxi) $x^2 + y^2 + 2x = 0$

4. Graphs.

Exercises 9
Graphs

Exercises 12

(i) $r = 5 \sec\left(\Theta - \dfrac{\pi}{2}\right)$

(ii) $r = 2 \sec\left(\Theta - \dfrac{\pi}{2}\right)$ at $\Theta = \dfrac{\pi}{4}$,

$r = -2 \sec\left(\Theta - \dfrac{\pi}{2}\right)$ at $\Theta = -\dfrac{\pi}{4}$

(iii) $r = \dfrac{15\sqrt{3}}{4} \sec\left(\Theta - \dfrac{\pi}{2}\right)$ at $\Theta = \dfrac{\pi}{3}$,

$r = \dfrac{-15\sqrt{3}}{4} \sec\left(\Theta - \dfrac{\pi}{2}\right)$ at $\Theta = -\dfrac{\pi}{3}$.

Exercises 13

(i) $r = \frac{5}{2} \sec \Theta$ at $\Theta = \frac{\pi}{4}$, $r = -\frac{5}{2} \sec \Theta$ at $\Theta = \frac{3\pi}{4}$

(ii) $r = \sec \Theta$ at $\Theta = \pi$, $r = 5 \sec \Theta$ at $\Theta = 0$, and $r = -\frac{1}{3} \sec \Theta$ at $109° 28'$ and $250° 32'$

(iii) $r = 8 \sec \Theta$ at $\Theta = 0$, $r = 0$ at $\Theta = \pi$ and $r = -\sec \Theta$ at $\Theta = \frac{2\pi}{3}$ and $\Theta = \frac{4\pi}{3}$

(iv) $r = 5 \sec \Theta$ at $\Theta = 0$, $r = -5 \sec \Theta$ at $\Theta = \pi$

$r = -1.36 \sec \Theta$ at $\Theta = 69° 18'$ and

$\Theta = -69° 18'$

$r = -1.36 \sec \Theta$ at $\Theta = 114° 06'$ and

$\Theta = -114° 06'$.

Exercises 14

1. (i) 0.441 sqaure units

(ii) 0.447 sqaure units

(iii) 0.441 sqaure units

(iv) 0.45 sqaure units

2. (i) 2π square units
   (ii) $\dfrac{25\pi}{8}$ square units
   (iii) $\dfrac{25\pi}{12}$ square units
   (iv) $\dfrac{3\pi}{4}$ square units

3. $\dfrac{3\pi^3}{8}$ square units

4. $\dfrac{25}{4}\left(\dfrac{\pi}{3}+\dfrac{\sqrt{3}}{2}\right)$ square units

5. 2π square units

6. $\dfrac{75\pi}{8}-25$ square units

7. 2.45 square units

8. 38.3 square units

9. 2 square units

10. (i) a^2 square units
    (ii) a^2 square units.

5. CARTESIAN AND POLAR CURVE SKETCHING

Index

A
Algebraic function
    cubic 4
    quadratic 3
Algebraic-exponential
    function 11–12
Algebraic with asymptotes 14
Areas of polar curves 43
Areas of tirangles 36
    polar co-ordinates
Asymptotes 14

C
Cartesian curve sketching
    cubic 4
    quadratic 3–4
Cartesian to polar 37
Composite functions 11
Cubic functions 4

E
Exercises
1   9
2   13
3   17
4   25
5   28
6   29
9   35
12   41

H
Half lines 31

I
Initial line 27

M
Maximum power transfer 6
Miscellaneous 45

P
Points of inflexion 4
Polar co-ordinate system 27
Polar curve sketching 33
Polar to cartesian 37
Polar graphs 38–9
Polar equation of straight line 32
Pole 27

Q
Quadratic functions 3

R
Reciprocal (Graphs) 20
Relationship between polar and
    cartesian co-ordinates 29

S
Sign convention of polar
    co-ordinate system 27
Simple Transformation 18
Solutions to Miscellaneous 49
Stationary value of
    $r \sin \Theta$ 40
    $r \cos \Theta$ 42
Stationary points 3

T
Transformations
    $f(x)$ to $f(kx)$ 18
    $f(x)$ to $f(x) + k$ 19
    $f(x)$ to $-f(x)$ 19
    $f(x)$ to $f(-x)$ 20
    $f(x)$ to $kf(x)$ 20
    $f(x)$ to $f(x - k)$ 19
    $f(x)$ to $\frac{1}{f(x)}$ 20–24
Turning points 3